Farming Based Livelihood System

NIPA® GENX ELECTRONIC RESOURCES & SOLUTIONS P. LTD.
New Delhi-110 034

About the Authors

Prof (Dr) A Zaman is an internationally renowned scientist specialization with agricultural water management having spectacular career of 46 years. His lifetime and original research contribution in the field of irrigation water management, rainfed agriculture, cropping system, watershed management, dry land agriculture, agricultural heritage and bio-diversity conservation, to name a few, globally appreciated leading to outstanding teachers in farm universities and agricultural extension services in India and abroad. A large number original concepts, strategies, techniques a tools those developed which have contributed enormously to agricultural science and benefitted the global society. He served as Irrigation Agronomist and formerly Dean, Faculty of Agriculture, Director of Research, Head, Department of Agronomy in Bidhan Chandra Krishi Viswavidyalaya [SAU], West Bengal, India. He is a former Chief Scientist in AICRP on Water Management (ICAR) and former Associate Director of Research, Regional Research Station, Old Alluvial Zone (Presently UBKV, Pundibari). He has coordinated, associated and involved with post-graduate teaching programme at the University and has significant research experiences in field of water management and in academia and well as having considerable administrative experiences in different capacity at various time frame. Dr. Zaman worked as Head, Irrigation Expatriate Team, Government of Uganda for 2-year period (1994-1996) under ITEC Programme being deputed by the Government of India to design, lay out and installation of famers managed small scale irrigation for sustainable crop cultivation. He was also designated as IFFCO-Chair Professor. His scholarly publications include more than 200 research papers most of which published in peer-reviewed journals having national and international reputes, 9 book chapters, 7 books with global publishing houses. His academic contribution was profusely appreciated by the International Institute of Bengal and Himalayan Basin patronized by Nobel laureate Charles Town of California, USA where he became co-founder and organized International Conferences in India and Bangladesh bi-annually. He organized and chaired many prestigious international workshops, chaired many technical sessions in scientific meetings and platforms. He has delivered number of invited plenary lectures and key-note addresses and acting as member of Editorial Board members, Examiner and question setter in various agricultural universities in India and Bangladesh. He has 179 publications in the proceedings of national and international seminars and conferences held in the country and abroad in his credit. He guided and supervised 7 (seven) Ph D and at least 50 (fifty) students towards post graduate thesis work. He also worked as Programme Coordinator, West Bengal Accelerated Development of Minor Irrigation Project (World Bank) from 20 September 2014 to 2nd September 2016 after superannuation from the University Service. Prof (Dr) Zaman worked as Emeritus Professor, M S Swaminathan School of Agriculture, CUTM, Paralakhemundi, Odisha since 3rd September 2016 to 31 May 2018.

He was acting as Professor & In charge, founded agricultural education in The Neotia University, Kolkata campus at Sarisa, West Bengal from May, 2018 to April, 2020, organized the School of Agricultural Science in the university and introduced the agricultural academic course curriculum in the university from the academic session of 2018-19 in the university. At present, working as the Dean, School of Agricultural Sciences, Sister Nivedita University, Kolkata being joined on and from 1 May 2020.

Dr. Md Hedayetullah, Ph.d. is an Assistant Professor (Stage II) in the Department of Agronomy and the Officer In-Charge of the All India Coordinated Research Project (AICRP) on Chickpea (ICAR-IIPR) at the Directorate of Research, Bidhan Chandra Krishi Viswavidyalaya (BCKV), Mohanpur, Nadia, West Bengal, India. Additionally, he serves as the Farm Administrator of AB Block Farm, BCKV, Kalyani.

Dr. Hedayetullah has extensive experience in agronomy and agricultural research. He previously worked as an Agronomist at the National Bank for Agriculture and Rural Development (NABARD), Balasore, Odisha, and as an Assistant Professor at the M.S. Swaminathan Institute of Agriculture Science, Centurion University of Technology and Management, Gajapati, Odisha. He also served as an Assistant Professor at the College of Agriculture, Government of Tripura, India.

He holds a Bachelor of Science (Agriculture) degree from H.N.B. Garhwal University, Uttarakhand, India, a Master of Science (Agronomy) from Palli Siksha Bhavana, Institute of Agriculture, Visva-Bharati University, Sriniketan, West Bengal, and a PhD (Agronomy) from Bidhan Chandra Krishi Viswavidyalaya (BCKV), West Bengal, India.

Dr. Hedayetullah is a recipient of the prestigious Maulana Azad National Fellowship Award from the University Grants Commission (UGC), New Delhi, India. He has also secured multiple research fellowships from various funding agencies. Recently, he was honored with the "Eminent Scientist

Award" by Brainware University, Kolkata, and ISAHRD, Chandigarh, at the 9th International Conference on ICANFTA-2025.

With an extensive research background, Dr. Hedayetullah has authored and co-authored 50 research papers, 5 review papers, 30 book chapters, and 9 books. He has successfully guided 8 postgraduate students and is currently supervising 3 PhD students. He has participated in over 30 national and international seminars, symposia, conferences, workshops, and training programs. He is actively involved as a Principal Investigator (PI) and Co-Principal Investigator (Co-PI) in several national and international funded research projects.

Dr. Hedayetullah is an Editorial Board Member of the Amity Journal of Agribusiness (AJAB) and a life member of several esteemed scientific organizations, including:

- Crop and Weed Science Society, BCKV, West Bengal, India
- Indian Society of Pulses Research and Development
- Fire Asuk Sobuj
- Academy of Advancement of Agricultural Sciences (AAAS)

He is also a member of the International Institute Bengal Basin, Berkeley, USA, and the Asian Council of Science Editors (ACSE), Toronto, Canada.

Dr. Hedayetullah has actively contributed to agricultural extension programs through participation in various institutional and Krishi Vigyan Kendra (KVK) activities, farmers' training programs, and Krishi Melas. He has delivered several live TV talks on "Krishi Darshan," telecasted by Doordarshan Kolkata. He has also organized numerous Front Line Demonstration (FLD) programs and Tribal Sub-Plan (TSP) FLD farmers' training programs under rice fallow systems, funded by ICAR-Indian Institute of Pulses Research, under AICRP on Chickpea.

Farming Based Livelihood System

A Zaman
Dean, School of Agricultural Sciences
Sister Nivedita University
Kolkata-700 156, West Bengal

Md. Hedayetullah
Assistant Professor (Stage-II)
Department of Agronomy
BCKV & Officer-in-Charge
AICRP on Chickpea (ICAR)
BCKV Mohanpur-741252, West Bengal

NIPA® GENX ELECTRONIC RESOURCES & SOLUTIONS P. LTD.
New Delhi-110 034

NIPA® GENX ELECTRONIC
RESOURCES & SOLUTIONS P. LTD.
101,103, Vikas Surya Plaza, CU Block
L.S.C. Market, Pitam Pura, New Delhi-110 034
Ph : +91-11-43860225, Mob.: +91 9717133558, 9540816132
E-mail: newindiapublishingagency@gmail.com
Website: www.nipaersources.com

Print ISBN: 978-93-58879-68-1
ebook ISBN: 978-93-58873-92-4

Composed and Designed by NIPA®.

Preface

In a rapidly evolving world, the significance of agriculture and sustainable farming practices cannot be overstated. The book "Farming Based Livelihood System" serves as a comprehensive exploration of how agricultural practices are intertwined with livelihoods, shaping not only the economies but also the cultural and social fabrics of communities worldwide. Rooted in the belief that farming is not merely an occupation but a way of life, this compilation of 28 chapters seeks to illuminate diverse aspects of farming systems and their critical role in sustaining human lives and the environment.

The journey through this book delves into various dimensions of farming as a livelihood source, encompassing traditional practices, modern innovations, and the socio-economic challenges faced by farmers today. From the intricacies of sustainable agriculture to the impacts of climate change on crop yields, we aim to provide a holistic understanding of how farming influences and is influenced by myriad factors—ranging from global markets to local ecosystems.

Each chapter has been meticulously crafted by experts and practitioners in the field, bringing together a wealth of knowledge and experience. Readers will encounter case studies, practical insights, and theoretical frameworks that highlight successful models of farming-based livelihoods. We will explore not only the agricultural techniques themselves but also the necessity for community engagement, policy support, and technological advancements in fostering resilience among farming communities.

As we navigate through topics such as agro-ecology, crop diversification, organic farming, and the role of cooperative societies, the reader will gain a nuanced perspective on the complexities of agricultural systems. Furthermore, the social implications of farming—addressing issues like land rights, food security, and gender equality—are examined to emphasize the importance of inclusivity in creating sustainable livelihoods.

This book is intended for a diverse audience: students, researchers, policymakers, and practitioners who share an interest in agriculture and its potential to uplift communities. It is our hope that "Farming Based Livelihood System" will inspire a deeper appreciation for the challenges and opportunities within the agricultural sector, encouraging informed discussions, innovative solutions, and committed action towards a more sustainable future.

As you embark on this exploration of farming systems, we invite you to reflect on the vital connections between the soil, the farmers, and the communities nourished by their hard work. May this book serve as a catalyst for change, fostering a renewed commitment to sustainable farming practices and the livelihoods they sustain.

Sincerely

Authors

Contents

Practical (Experiments)

Farm Based Livelihood System

1

Status of Agriculture in India and Different States

What is a Farming-Based Livelihood System?

At its core, a farming-based livelihood system is one where agriculture (crop production, livestock rearing, or a combination) forms the *primary* source of income, food security, and overall well-being for a household or community. It's more than just growing food; it's an interconnected system involving:

- **Production Activities:** This includes all the activities directly related to farming, like:
  - Crop cultivation (e.g., cereals, fruits, vegetables, cash crops)
  - Livestock rearing (e.g., cattle, poultry, goats, pigs, aquaculture)
  - Agroforestry (integrating trees and shrubs into farming systems)
  - Soil and water management
- **Assets:** These are the resources available to the farming household. Key asset categories are often described as:
  - *Natural Capital:* Land, water, soil quality, biodiversity, climate.
  - *Physical Capital:* Infrastructure (irrigation, roads, storage facilities), tools, equipment, livestock.
  - *Human Capital:* Skills, knowledge, labor availability, health, education.
  - *Financial Capital:* Savings, access to credit, insurance.
  - *Social Capital:* Networks, community organizations, access to information, trust.
- **Livelihood Strategies:** These are the ways that farming households combine their assets to generate income and meet their needs. Strategies can include:
  - *Intensification:* Increasing production per unit of land (e.g., using fertilizers, improved seeds, irrigation).

  - *Diversification:* Engaging in a wider range of agricultural activities (e.g., growing multiple crops, integrating livestock). Also, diversifying into *non-farm* activities is a key element of many rural livelihoods.
  - *Migration:* Seeking employment elsewhere (seasonal or permanent).
  - *Specialization:* Focusing on a particular crop or livestock product.
- **Vulnerability Context:** External factors that can affect the livelihood system, such as:
  - Climate change (droughts, floods, extreme weather events).
  - Market fluctuations (price volatility).
  - Pests and diseases.
  - Policy changes.
  - Political instability.
- **Policies and Institutions:** Government policies, laws, and institutions (e.g., extension services, research organizations, land tenure systems) that influence farming practices and outcomes.
- **Livelihood Outcomes:** The results of the livelihood strategies, such as:
  - Increased income.
  - Improved food security.
  - Better health and nutrition.
  - Enhanced social well-being.
  - Sustainable resource management.

Key Characteristics of Farming-Based Livelihoods:

- **Dependence on Natural Resources:** Farming is inherently tied to the environment. The quality of land, water availability, and climate patterns significantly impact productivity.
- **Seasonality:** Agricultural production often follows seasonal cycles, leading to fluctuations in income and labor demand.
- **Vulnerability to Risk:** Farmers face numerous risks, including weather-related disasters, pest outbreaks, and market volatility.
- **Labor-Intensive:** Many farming systems, especially in developing countries, rely heavily on manual labor.
- **Close Link to Food Security:** Farming households often consume a significant portion of what they produce, making agriculture crucial for their food security.

- **Cultural Significance:** Farming is often deeply embedded in cultural traditions and social structures.

Types of Farming Systems

Farming-based livelihood systems are incredibly diverse. Here are a few broad classifications:

- **Subsistence Farming:** Primarily focused on producing enough food to meet the needs of the household. Often characterized by low input use and limited surplus for sale.
- **Commercial Farming:** Focused on producing crops or livestock for sale in the market. Often involves higher levels of technology and input use.
- **Mixed Farming:** Integration of crop and livestock production. This can provide diversification and improve resource utilization.
- **Intensive Farming:** High input (e.g., fertilizers, pesticides, irrigation) and high output per unit of land.
- **Extensive Farming:** Low input and low output per unit of land. Often practiced in areas with limited resources.
- **Organic Farming:** Focuses on sustainable practices, avoiding synthetic inputs.
- **Agroforestry:** Integrating trees and shrubs into farming systems to provide multiple benefits (e.g., timber, fruits, shade, soil conservation).
- **Pastoralism:** Livestock rearing is the primary activity, often involving nomadic or semi-nomadic grazing patterns.

Challenges and Opportunities

Farming-based livelihoods face numerous challenges:

- **Climate Change:** Shifting weather patterns, increased frequency of extreme events.
- **Land Degradation:** Soil erosion, nutrient depletion, deforestation.
- **Water Scarcity:** Competition for water resources, inefficient irrigation practices.
- **Market Access:** Limited access to markets, low prices, lack of infrastructure.
- **Lack of Technology:** Limited access to improved seeds, fertilizers, and other technologies.
- **Policy Constraints:** Unfavorable policies, lack of support for smallholder farmers.

- However, there are also significant opportunities:
- **Sustainable Intensification:** Increasing productivity while minimizing environmental impacts.
- **Climate-Smart Agriculture:** Adopting practices that reduce greenhouse gas emissions and enhance resilience to climate change.
- **Value Chain Development:** Improving market access and increasing the value of agricultural products.
- **Diversification:** Expanding into non-farm activities to supplement income.
- **Technology Adoption:** Using information and communication technologies (ICTs) to improve access to information and markets.
- **Policy Support:** Creating a supportive policy environment for smallholder farmers.

Important Considerations

- **Context Matters:** Farming systems are highly context-specific. Solutions need to be tailored to the local environment, culture, and economy.
- **Gender:** Women play a crucial role in agriculture, but often face significant constraints. Addressing gender inequality is essential for improving farming-based livelihoods.
- **Sustainability:** It's crucial to promote farming practices that are environmentally sustainable and ensure the long-term viability of agricultural production.
- **Integrated Approach:** Addressing the challenges of farming-based livelihoods requires an integrated approach that considers all aspects of the system, from production to markets to policies.
- **In summary, farming-based livelihood systems are complex and dynamic. Understanding the key components and challenges is essential for developing effective strategies to improve the well-being of farming households and communities.** Let me know if you would like me to elaborate on any of these aspects!

Status of Agriculture in India

Agriculture is the backbone of India's economy, accounting for about 18% of the country's GDP and providing employment to more than 50% of the population. However, the sector is facing numerous challenges, including:

1. **Low Productivity:** India's agricultural productivity is lower than that of many other countries, primarily due to inadequate irrigation, lack of mechanization, and outdated farming practices.
2. **Monsoon Dependence:** India's agricultural sector is heavily dependent on monsoon rains, which can lead to crop failures and droughts, affecting the overall productivity and income of farmers.
3. **Water Scarcity:** Water scarcity is a major issue in many parts of India, particularly during the summer months, affecting crop growth and productivity.
4. **Soil Degradation:** Soil degradation is a significant problem in India, primarily due to intensive farming practices, leading to soil erosion, nutrient depletion, and reduced fertility.
5. **Farm Distress:** Many farmers in India are facing financial difficulties, with high debt levels, low incomes, and limited access to credit and markets.

Status of Agriculture in India and Different States

Agriculture is the primary livelihood source for nearly half of India's population. It plays a key role in food security, rural employment, and the country's economy. India is among the top producers of cereals, pulses, fruits, vegetables, and spices. Agriculture in India is diverse, dynamic, and critical for inclusive growth. While several states have made strides in productivity and sustainability, others still face challenges due to climate and infrastructural constraints. Future growth depends on technology, market reforms, water management, and farmer empowerment.

Agriculture holds a pivotal place in India's socio-economic fabric. As a sector, it employs nearly 50% of the country's workforce and contributes approximately 18% to the national Gross Domestic Product (GDP). Though the share of agriculture in the overall economy has declined over the decades due to rapid industrialization and growth in the services sector, it remains crucial for food security, rural development, and poverty alleviation. India has made significant strides in agricultural production and is today one of the largest producers of rice, wheat, pulses, milk, fruits, and vegetables globally. India's agricultural landscape is marked by diversity in climatic zones, soil types, cropping patterns, and irrigation facilities. The country practices a wide range of agricultural systems—from intensive rice cultivation in the east to dryland farming in the arid west, and from plantation crops in the south to hill agriculture in the north. Such diversity not only supports food grain production but also enables the cultivation of a variety of horticultural crops, spices, oilseeds, and commercial crops like cotton and sugarcane.

Despite progress, Indian agriculture faces several structural and contemporary challenges. These include fragmentation of landholdings, declining soil fertility, overexploitation of groundwater, post-harvest losses, price volatility, and climate variability. The majority of Indian farmers are small and marginal, with limited access to credit, technology, and markets. Yet, in recent years, there has been a positive shift with increased focus on digital agriculture, farmer producer organizations (FPOs), climate-smart practices, and agri-tech innovations. The performance and characteristics of agriculture vary widely across Indian states due to differences in natural resources, policy support, and infrastructure. Punjab and Haryana, often referred to as the granaries of India, have highly mechanized and irrigated farming systems, especially for wheat and rice. However, they now face ecological concerns such as soil degradation and depleting groundwater levels due to decades of intensive monoculture.

Uttar Pradesh, the most populous state, leads in food grain production and is a major grower of sugarcane, wheat, and pulses. Madhya Pradesh has emerged as a leading state for pulses and soybean production, aided by its improving irrigation infrastructure. Maharashtra has a diversified cropping system that includes cotton, sugarcane, and fruits, but also grapples with droughts and farmer distress in its semi-arid regions like Vidarbha and Marathwada. In eastern India, states like Bihar and West Bengal boast of fertile lands and high rainfall, making them suitable for rice, maize, and vegetables. However, these states often suffer from underdeveloped irrigation systems and inadequate access to post-harvest facilities. Odisha, with its tribal farming population, is pushing for agricultural diversification with millets and minor crops.

Rajasthan represents a classic case of arid and semi-arid agriculture, with a focus on coarse cereals like bajra and oilseeds like mustard. Gujarat, known for its entrepreneurial farmers, has made remarkable progress in cotton, groundnut, and dairy sectors. The success of the Amul cooperative model is a testament to the state's cooperative strength. Southern states like Andhra Pradesh, Telangana, Karnataka, and Tamil Nadu showcase varied farming systems—from rice and cotton to horticulture and spices. Andhra Pradesh has made significant strides in aquaculture and chilli production, while Telangana's irrigation schemes have boosted rice and maize yields. Karnataka, known for its ragi and pulses, is also emerging as a hub for organic farming and agri-startups. Tamil Nadu's traditional tank irrigation system and diverse cropping pattern make it a leading state in rice, banana, and sugarcane cultivation. In the northeast, states like Assam and others grow rice, tea, and mustard under high rainfall conditions, though agricultural development is often limited by poor connectivity and terrain-related challenges. Kerala, on the other hand, is known for plantation crops like coconut, rubber, and spices, and relies heavily on remittance income alongside farming.

Across the country, the government has launched several initiatives to strengthen agriculture, such as the PM-KISAN scheme for income support, the Soil Health Card scheme to promote balanced fertilizer use, and e-NAM to create a unified national market for agricultural produce. Programs like the Pradhan Mantri Krishi Sinchayee Yojana aim to enhance irrigation coverage and water-use efficiency. In recent years, a new wave of modernization is reshaping Indian agriculture. Technologies like drones, artificial intelligence, mobile-based advisory services, and smart irrigation are reaching the grassroots through startups and extension networks. Moreover, policies are increasingly aimed at improving value chains, supporting farmers' collectives, and boosting exports. A region-specific, climate-resilient, and technology-driven approach is the key to transforming Indian agriculture into a vibrant, inclusive, and sustainable sector.

- **Share in GDP (2023-24):** ~18%
- **Employment:** ~50% of the population
- **Net Sown Area:** ~140 million hectares
- **Major Crop Seasons:** Kharif (June–Oct), Rabi (Oct–March), Zaid (Summer)

All-India Agriculture Snapshot (2023-24)

Parameter	Data/Status
Net Sown Area	~140 million hectares
Gross Cropped Area	~195 million hectares
Irrigated Area	~48% of total cropped area
Total Food grain Production	~320 million tonnes
Major Crops	Rice, Wheat, Maize, Pulses, Cotton
Leading Exports	Basmati rice, spices, cotton, tea
Fertilizer Use	~140 kg/ha (national average)
Dominant Farming Type	Small & marginal holdings (85%)

Agriculture by State: Key Highlights

State	Major Crops	Features
Punjab	Wheat, Rice, Maize	Mechanized, irrigated, high yield, but groundwater stress
Haryana	Wheat, Rice, Sugarcane	Similar to Punjab; faces sustainability concerns
Uttar Pradesh	Wheat, Rice, Sugarcane, Pulses	Largest food grain producer; needs diversification
Madhya Pradesh	Soybean, Wheat, Pulses, Oilseeds	Leading in pulses; increasing irrigation

Maharashtra	Cotton, Sugarcane, Pulses	Rainfed, farmer distress in dry regions; strong horticulture
Bihar	Rice, Maize, Wheat, Pulses	Fertile land but underdeveloped irrigation and infrastructure
West Bengal	Rice, Jute, Vegetables	High productivity in small holdings; rich in fish and vegetables
Odisha	Rice, Pulses, Millets	Rainfed; tribal agriculture; millet promotion programs
Rajasthan	Bajra, Mustard, Maize	Arid/semi-arid; strong livestock and oilseed production
Gujarat	Cotton, Groundnut, Cumin	Strong cooperative sector (e.g., Amul), diverse cropping
Andhra Pradesh	Rice, Cotton, Chilies, Groundnut	Well-irrigated areas; aquaculture development
Telangana	Cotton, Rice, Maize	Growth in irrigation (Mission Bhagiratha); innovation in input subsidies
Karnataka	Ragi, Sugarcane, Pulses, Coffee	Horticulture and rainfed agriculture dominate
Tamil Nadu	Rice, Sugarcane, Banana	Intensive tank irrigation; active organic farming initiatives
Assam & NE States	Rice, Mustard, Tea	High rainfall; tea estates; need for better infrastructure
Kerala	Coconut, Spices, Rubber	Plantation crops; smallholder-dominated system

Government Initiatives

Scheme/Program	**Objective**
PM-KISAN	Direct income support to farmers
Soil Health Card Scheme	Promote balanced fertilizer use
Pradhan Mantri Krishi Sinchayee Yojana (PMKSY)	Improve irrigation efficiency
E-NAM (National Agriculture Market)	Unified online market for agricultural produce
National Food Security Mission (NFSM)	Boost pulse and foodgrain production
RKVY-RAFTAAR	Infrastructure and innovation in agri sector

Emerging Trends in Indian Agriculture

- **Digital Farming:** Use of drones, AI, IoT, mobile apps
- **Organic and Natural Farming**
- **Agri Startups:** In farm inputs, processing, logistics
- **Farmer Producer Organizations (FPOs):** Collectivizing small farmers
- **Climate-Smart Agriculture:** Promoting resilient and sustainable practices

QUESTION BANK

Short-answer questions

1. What is the contribution of agriculture to India's GDP?
2. Which state is the largest producer of wheat in India?
3. Name any two major cash crops grown in India.
4. What is the primary farming system practiced in the northeastern states of India?
5. Which state is known as the "Rice Bowl of India"?
6. What percentage of India's population depends on agriculture for livelihood?
7. Name the first Indian state to become fully organic in farming.
8. Which Indian state leads in milk production?
9. What is the major cropping pattern in the Indo-Gangetic Plains?
10. What is the objective of the National Agriculture Market (eNAM)?

Multiple Choice Questions (MCQs)

1. What is the contribution of agriculture to India's GDP as of recent estimates?

   a) 10-12% b) 15-17%
   c) 20-25% d) 30-35%

2. Which state is the largest producer of wheat in India?

   a) Punjab b) Uttar Pradesh
   c) Madhya Pradesh d) Haryana

3. Which of the following farming systems is most prevalent in the northeastern states of India?

   a) Commercial farming b) Plantation farming
   c) Jhum (shifting) cultivation d) Horticulture-based farming

4. Which state is known as the "Rice Bowl of India"?

   a) West Bengal b) Punjab
   c) Andhra Pradesh d) Chhattisgarh

5. What percentage of India's population is directly or indirectly dependent on agriculture?

   a) 30% b) 45%
   c) 50% d) 60%

6. Which state leads in organic farming practices in India?
   a) Punjab b) Sikkim
   c) Maharashtra d) Kerala
7. The Green Revolution primarily benefited which region of India?
   a) North-Eastern states b) Western Ghats
   c) Indo-Gangetic Plains d) Deccan Plateau
8. What is the major cropping pattern followed in the Indo-Gangetic Plains?
   a) Millets-Pulses b) Wheat-Rice
   c) Sugarcane-Maize d) Cotton-Groundnut
9. Which Indian state is the largest producer of pulses?
   a) Maharashtra b) Madhya Pradesh
   c) Rajasthan d) Uttar Pradesh
10. Which is the most cultivated crop in India in terms of area?
   a) Wheat b) Rice
   c) Maize d) Sugarcane

True/False Questions

1. India is the world's largest producer of spices. **(True)**
2. Agriculture in India is primarily dependent on monsoon rainfall. **(True)**
3. Punjab is the leading producer of rice in India. **(False)** *(West Bengal is the largest producer)*
4. India ranks first in global milk production. **(True)**
5. Jhum cultivation is widely practiced in Tamil Nadu. **(False)** *(It is practiced in North-Eastern states)*
6. Maharashtra is the leading producer of sugarcane in India. **(False)** *(Uttar Pradesh is the leading producer)*
7. The livestock sector contributes significantly to farmers' income in India. **(True)**
8. West Bengal is the leading producer of potatoes in India. **(True)**
9. The National Agriculture Market (eNAM) is an online trading platform for agricultural commodities. **(True)**
10. India has achieved self-sufficiency in food grain production. **(True)**

2

Income of Farmers and Rural People in India

India's rural economy plays a vital role in the overall development of the country, with a significant portion of the population dependent on agriculture and allied sectors for their livelihood. Despite the structural changes in the Indian economy over the past decades, agriculture continues to be the primary source of income for a majority of rural households. However, the income levels of farmers and rural people vary widely due to factors such as landholding size, access to resources, market linkages, climatic conditions, education, and government support systems.

The average monthly income of agricultural households in India has been a matter of concern and active policy intervention. According to reports from the National Sample Survey Office (NSSO) and other government bodies, many small and marginal farmers earn less than what is required for a sustainable livelihood. This income typically comes from multiple sources, including crop production, livestock rearing, wages, and non-farm activities. Yet, the increasing input costs, fluctuating market prices, and the vulnerability to climatic shocks make farming a risky and often unremunerative occupation. To address these challenges, various initiatives have been undertaken by the Government of India, such as the **Doubling Farmers' Income (DFI)** strategy, minimum support prices (MSP), crop insurance schemes like PMFBY (Pradhan Mantri Fasal Bima Yojana), rural employment guarantee schemes like MGNREGS, and efforts to promote farmer producer organizations (FPOs). Despite these measures, the rural income landscape remains uneven, and regional disparities persist across different states and districts. Non-farm rural employment also contributes significantly to rural incomes. With the growth of rural infrastructure, skill development programs, and better connectivity, rural youth are increasingly finding employment in construction, retail, transportation, and small-scale industries. Remittances from rural-to-urban migrants also form an important part of rural household income in many areas.

Improving the income of farmers and rural people in India requires an integrated approach involving increased investment in agriculture, better access to credit and technology, market reforms, and targeted social welfare programs. Empowering women, encouraging agri-entrepreneurship, promoting sustainable farming practices, and enhancing rural education and healthcare are also critical to ensuring inclusive rural development. The income of farmers and rural people in India is both a reflection of the country's socio-economic conditions and a key determinant of its future growth. Bridging the rural-urban income gap is essential not just for reducing poverty but also for achieving equitable and sustainable development across the nation.

According to the **NABARD All India Rural Financial Inclusion Survey (NAFIS)** and the **National Sample Survey Office (NSSO)** data, the average monthly income of an agricultural household in India was estimated to be around □10,218 in 2018-19. This income comes from a mix of sources—cultivation, livestock, wages, and non-farm activities. However, there is considerable variation between states, with farmers in Punjab, Haryana, and Kerala earning significantly more than those in states like Bihar, Odisha, and Jharkhand.

Sources of Income in Rural Areas

The income of rural people in India generally comes from multiple sources, including:

1. **Agriculture (Crop Production):** The largest source of income for rural households, especially among small and marginal farmers.
2. **Livestock Rearing:** A critical supplementary income source, particularly in states like Rajasthan, Gujarat, and Uttar Pradesh.
3. **Wage Labour:** Many rural households, especially landless ones, depend on casual wage labor in agriculture or construction.
4. **Non-Farm Activities:** These include petty trade, handicrafts, services (like driving, tailoring), and rural entrepreneurship.
5. **Remittances:** Migration to urban areas for employment contributes to rural household income through remittances.

Challenges in Enhancing Farm Income

Several structural and operational issues hinder the growth of farmer income:

- **Fragmented Landholdings:** Around 85% of Indian farmers own less than 2 hectares, making it difficult to generate adequate income from farming alone.
- **Low Productivity:** Poor access to quality inputs, irrigation, and technology results in lower yields.

- **Market Risks:** Farmers often face price fluctuations and lack bargaining power in markets.
- **Post-Harvest Losses:** Due to inadequate storage and processing facilities.
- **Debt Trap:** Many farmers borrow from informal sources at high interest rates, which affects their net earnings.

Income of Farmers in India

There exists a wide disparity in income levels across different states. Farmers in **Punjab, Haryana, and Maharashtra** generally have higher incomes due to better irrigation, infrastructure, and access to markets. In contrast, **Bihar, Uttar Pradesh, Jharkhand, and parts of the northeast** continue to struggle with low farm incomes due to smaller landholdings, poor irrigation, and limited non-farm opportunities. The average income of a farmer in India is around □50,000-60,000 (approximately $665-$800 USD) per year. However, this figure varies widely depending on the type of crop, region, and other factors. The average income of rural people in India is around □20,000-30,000 (approximately $265-$400 USD) per month. However, this figure can vary depending on factors such as occupation, education, and access to resources.

State-wise Agricultural Performance

Here's a brief snapshot of the agricultural performance of various states in India:

1. **Punjab:** Known for its high productivity and yields of wheat and rice, Punjab is one of the most prosperous agricultural states in India.
2. **Haryana:** Haryana is another major agricultural producer in India, with a focus on crops like wheat, rice, and sugarcane.
3. **Uttar Pradesh:** Uttar Pradesh is the largest producer of wheat and maize in India, with a significant contribution to the country's agricultural economy.
4. **Madhya Pradesh:** Madhya Pradesh is one of the largest producers of soybean and corn in India, with a growing focus on oilseeds and other commercial crops.
5. **Maharashtra:** Maharashtra is a major producer of sugarcane, cotton, and soybean, contributing significantly to India's agricultural economy.
6. **Karnataka:** Karnataka is a significant producer of coffee, sugarcane, and tobacco, with a growing focus on horticultural crops like fruits and vegetables.

7. **Andhra Pradesh:** Andhra Pradesh is known for its high productivity and yields of rice, with a focus on horticultural crops like mangoes and citrus fruits.
8. **Tamil Nadu:** Tamil Nadu is a major producer of cotton, sugarcane, and groundnuts, with a growing focus on horticultural crops like fruits and vegetables.
9. **West Bengal:** West Bengal is a significant producer of rice and jute, with a growing focus on horticultural crops like mangoes and citrus fruits.

Initiatives to Improve Agricultural Performance

The Indian government has launched several initiatives to improve agricultural performance and support farmers, including:

1. **Pradhan Mantri Kisan Samman Nidhi (PM-KISAN):** A scheme to provide financial assistance to farmers, with a focus on small and marginal farmers.
2. **Soil Health Card Scheme:** A scheme to provide soil health cards to farmers, with recommendations for fertilizer applications and crop management.
3. **National Mission for Sustainable Agriculture (NMSA):** A mission to promote sustainable agriculture practices, including organic farming and agroforestry.
4. **National Horticulture Mission (NHM):** A mission to promote horticulture and support farmers engaged in fruit and vegetable cultivation.
5. **E-NAM (Electronic National Agriculture Market):** A platform to connect farmers with buyers and enable online trading of agricultural commodities.
6. **Minimum Support Price (MSP):** Ensures remunerative prices for major crops, though accessibility varies.
7. **Fasal Bima Yojana:** Offers crop insurance to protect against yield loss due to weather and pests.
8. **Rural Livelihood Missions (NRLM):** Empower women and self-help groups through skill development and micro-enterprises.
9. **MGNREGA (Mahatma Gandhi National Rural Employment Guarantee Act):** Guarantees 100 days of wage employment per year for rural households, enhancing non-farm rural income.

These initiatives aim to address the challenges facing India's agricultural sector, improve farmer income, and promote sustainable agriculture practices. However, the success of these initiatives will depend on the effective implementation and scale-up of these programs, as well as the availability of adequate resources and support systems.

Emerging Trends in Rural Income Generation

- **Agri-Entrepreneurship:** Youth in rural areas are now exploring agri-based startups, processing units, and service centers.
- **Digital Platforms:** Apps and digital kiosks help farmers access markets, weather forecasts, and government schemes.
- **FPOs (Farmer Producer Organizations):** Enable collective marketing and better negotiation power.
- **Diversification:** Many rural families are diversifying into dairying, poultry, fisheries, and horticulture for higher income.

QUESTION BANK

Short Answer Questions

1. What is the average monthly income of an Indian farmer as per recent government data?
2. Name two major sources of non-farm income for rural households.
3. Which state in India has the highest average farmer income?
4. What is the primary occupation of rural people apart from farming?
5. How does the Pradhan Mantri Kisan Samman Nidhi (PM-KISAN) scheme support farmers?
6. Name one key factor affecting the income of farmers in India.
7. What percentage of rural households in India are engaged in agriculture?
8. How does Minimum Support Price (MSP) impact farmer income?
9. What is the role of rural self-employment programs like MGNREGA?
10. Name one cash crop that significantly boosts farmer income in India.

Multiple Choice Questions (MCQs)

1. What is the major source of income for farmers in India?

   a) Government jobs b) Crop cultivation

   c) Real estate d) Factory work

2. Which scheme provides direct financial assistance to small and marginal farmers?

   a) PM-KISAN b) PMAY

   c) Ayushman Bharat d) Make in India

3. What is the biggest factor affecting rural household income?

   a) Industrialization b) Agricultural productivity

   c) Tourism d) Social media

4. Which state has the lowest average farmer income in India?

   a) Bihar b) Punjab

   c) Kerala d) Haryana

5. What is the primary objective of MGNREGA?

   a) Provide urban employment

   b) Offer free education

   c) Ensure minimum wage rural employment

   d) Promote digital transactions

6. Which farming method is known to improve income sustainability for farmers?

   a) Monocropping b) Organic farming

   c) Slash-and-burn farming d) Jhum cultivation

7. Which of the following is NOT a major source of rural income?

   a) Agriculture b) Livestock farming

   c) Stock market trading d) Handicrafts

8. How often is the Minimum Support Price (MSP) revised in India?

   a) Annually b) Every 5 years

   c) Every 2 years d) Monthly

9. Which crop provides the highest return per hectare in India?

   a) Rice b) Sugarcane

   c) Cotton d) Tea

10. Which factor contributes most to fluctuating farmer income in India?

    a) Government policies b) Monsoon dependency

    c) Availability of labor d) Road infrastructure

True/False Questions

1. Agriculture is the primary source of income for over 50% of India's population. **(True)**
2. The average farmer's income in India is higher than that of urban workers. **(False)**
3. Livestock farming is a significant contributor to rural household income. **(True)**
4. PM-KISAN provides □10,000 per year to small and marginal farmers. . **(False)** *(It provides □6,000 annually)*
5. Minimum Support Price (MSP) is fixed by state governments.
6. Rural employment guarantee schemes help in increasing farmer income. **(True)**
7. Climate change has no effect on farmer income in India. **(False)**
8. Cash crops generally provide higher income than food grains. **(True)**
9. Small farmers in India mostly depend on rain-fed agriculture. **(True)**
10. Direct farm-to-market sales always result in lower income for farmers. **(False)**

3

Livelihood- Definition, Concept and Livelihood Pattern in Urban and Rural Areas

Livelihood refers to the means and capabilities by which individuals and households secure the necessities of life such as food, shelter, clothing, education, and healthcare. It encompasses not just the income-generating activities people undertake, but also the assets, skills, knowledge, and social networks they use to sustain and improve their living conditions. The concept of livelihood is broader than employment or income; it includes access to natural, financial, social, human, and physical capital that enable people to build a decent standard of living.

A widely accepted definition by the Sustainable Livelihood Framework (developed by the UK's Department for International Development) states that a livelihood comprises "the capabilities, assets (including both material and social resources), and activities required for a means of living." A livelihood is considered sustainable when it can cope with and recover from stresses and shocks, maintain or enhance its capabilities and assets, and provide sustainable livelihood opportunities for the next generation. In India, livelihood patterns differ significantly between urban and rural areas due to differences in economic structures, resource availability, access to services, and lifestyle choices.

In **rural areas**, livelihoods are primarily dependent on **agriculture and allied activities** such as animal husbandry, forestry, and fishing. Most rural households engage in subsistence or small-scale farming, often with limited access to land, irrigation, modern technology, and markets. Seasonal migration to nearby towns and cities for wage labor is also a common livelihood strategy among rural poor, especially during lean agricultural seasons. In addition to farming, non-farm activities like pottery, weaving, carpentry, and running small shops also support rural incomes. The **urban livelihood pattern**, on the other hand, is more diversified and generally oriented towards **non-agricultural sectors**. People in urban areas find employment in industry, trade, construction, transportation, services, and informal sectors such as street vending, domestic work, and waste recycling. Urban livelihoods are more influenced by market demands, educational qualifications, and skill levels. While cities offer more opportunities for income generation, they also present challenges like job insecurity, informal employment, poor working conditions, and high living costs, especially for the urban poor.

It is important to recognize that both urban and rural livelihoods are vulnerable to external shocks such as economic downturns, environmental changes, health emergencies, and policy shifts. Rural livelihoods may be disrupted by droughts or floods, while urban livelihoods can be affected by industrial closures or housing evictions. Thus, promoting resilient and sustainable livelihoods is a central goal of development planning. Efforts to strengthen livelihoods in both contexts include skill development programs, access to microfinance, promotion of self-help groups, digital literacy, rural employment schemes like MGNREGA, urban livelihood missions such as DAY-NULM, and social protection initiatives. A livelihood-centered approach to development ensures that people not only have jobs or incomes but also the dignity, agency, and security they need for a meaningful life. In essence, the concept of livelihood provides a holistic framework to understand how people make a living in diverse and dynamic environments. Whether in a rural village or an urban slum, securing a sustainable livelihood remains at the heart of human development.

Definition of Livelihood

A livelihood encompasses the capabilities, assets (including both material and social resources), and activities required for a means of living. It is sustainable when it can cope with and recover from stresses and shocks maintain or enhance its capabilities and assets, while not undermining the natural resource base. In simpler terms:

- **Capabilities:** The skills, knowledge, and health that people possess.
- **Assets:** The resources people have access to, like land, equipment, savings, social networks, etc.
- **Activities:** What people do to generate income or obtain necessities (farming, wage labour, trade, etc.)?

Key Concepts of Livelihoods

- **Sustainability:** A key aspect. A livelihood isn't just about making money *now*, but ensuring future generations can also meet their needs. This involves protecting the environment and building resilience.
- **Vulnerability:** Livelihoods are often vulnerable to shocks (e.g., economic downturns, natural disasters, and illness) and stresses (e.g.,

increasing competition, environmental degradation). Understanding and addressing vulnerability is crucial.

- **Assets-Based Approach:** Focuses on what people *have* and how they can build upon it, rather than just focusing on their needs or deficits.
- **Holistic:** Livelihoods are interconnected. Economic, social, environmental, and political factors all play a role.
- **Agency:** Recognizes that people are not passive victims, but active agents in shaping their own livelihoods.

Livelihood Patterns in Urban Areas

Urban areas are characterized by higher population densities, greater economic diversification, and more developed infrastructure compared to rural areas. Here are some common livelihood patterns:

- **Formal Sector Employment**
  - **White-Collar Jobs:** Professional, managerial, and administrative roles in offices, government, and large organizations. → implies → higher salaries, benefits, and job security (though this can vary).
  - **Blue-Collar Jobs:** Skilled and unskilled manual labour in factories, construction, transportation, and other industries. → \implies → more physically demanding, potentially lower pay and less job security than white-collar jobs.
- **Informal Sector Employment**
  - **Small-Scale Trading:** Street vendors, market stalls, and small shops selling goods and services. → \implies → low barriers to entry but often low and unstable income.
  - **Service Provision:** Domestic workers, construction labourers, rickshaw pullers, auto-rickshaw drivers, etc. → \implies → often precarious, with long hours, low wages, and limited legal protection.
  - **Waste Picking/Recycling:** Collecting and sorting recyclable materials from waste. → \implies → very low income, hazardous working conditions, and social stigma.
- **Entrepreneurship**
  - **Small Businesses:** Restaurants, cafes, retail stores, workshops, and service businesses. → \implies → potential for higher income but also involves risk and requires capital and business skills.
  - **Technology-Based Businesses:** Start-ups in software, e-commerce, and other digital sectors. → \implies → high growth potential but also highly competitive and requires specialized skills.

- **Remittances:** Urban residents often send money back to their families in rural areas. This can be a significant source of income for rural households.

Key Characteristics of Urban Livelihoods

- **Monetization:** Greater reliance on cash income for food, housing, and other necessities.
- **Specialization:** More specialized occupations and industries.
- **Competition:** Higher levels of competition for jobs and resources.
- **Access to Services:** Potentially better access to education, healthcare, and infrastructure.
- **Vulnerability:** Vulnerable to economic fluctuations, job losses, and high cost of living.

Livelihood Patterns in Rural Areas

Rural areas are characterized by lower population densities, a greater reliance on natural resources, and often limited access to services and infrastructure. Common livelihood patterns include:

- Agriculture
  - **Subsistence Farming:** Producing food primarily for household consumption. → \implies → provides food security but often low income and vulnerable to weather and pests.
  - **Commercial Farming:** Producing crops or livestock for sale in the market. → \implies → potential for higher income but requires access to land, capital, and markets.
- **Livestock Keeping**
  - **Pastoralism:** Raising livestock for milk, meat, and other products, often involving nomadic or semi-nomadic lifestyles. → \implies → adapted to arid and semi-arid environments but vulnerable to drought and land degradation.
  - **Mixed Farming:** Integrating livestock with crop production. →\implies → diversifies income and provides manure for fertilizer.
- **Fishing**
  - **Artisanal Fishing:** Small-scale fishing using traditional methods. →\implies→ provides food and income for coastal communities but vulnerable to overfishing and environmental degradation.

  - **Commercial Fishing:** Large-scale fishing using modern equipment. → \implies→ higher catches but can deplete fish stocks and damage marine ecosystems.
- Forestry:
  - **Timber Extraction:** Harvesting timber for construction and other uses. →\ implies → can generate income but can also lead to deforestation and habitat loss.
  - **Non-Timber Forest Products (NTFPs):** Collecting fruits, nuts, mushrooms, medicinal plants, and other products from forests. → \ implies→ provides a supplementary source of income and food, often important for marginalized communities.
- **Rural Non-Farm Activities**
  - **Handicrafts:** Producing and selling crafts such as pottery, weaving, and woodwork. → \implies→ provides income and preserves traditional skills.
  - **Agro-Processing:** Processing agricultural products such as milling grain, making cheese, or canning fruits. → \implies→ adds value to agricultural products and creates employment.
  - **Tourism:** Providing services to tourists, such as accommodation, guiding, and selling souvenirs. → \implies→ generates income and promotes local culture.
- **Wage Labour**
  - **Agricultural Labour:** Working on other people's farms for wages. → \ implies→ often seasonal and low-paid.
  - **Construction Labour:** Working on construction projects in rural areas. → \implies→ physically demanding and often precarious.
- **Key Characteristics of Rural Livelihoods**
- **Dependence on Natural Resources:** Livelihoods are closely tied to land, water, forests, and other natural resources.
- **Seasonality:** Many rural activities are seasonal, leading to fluctuations in income and employment.
- **Limited Access:** Often limited access to markets, credit, education, healthcare, and other services.
- **Vulnerability:** Highly vulnerable to weather events, pests, diseases, and market fluctuations.
- **Social Capital:** Strong social networks and traditional institutions often play a crucial role in livelihood security.

Livelihoods are complex and context-specific. Understanding the different assets, activities, and vulnerabilities that shape people's livelihoods in urban and rural areas is essential for developing effective policies and interventions to improve their well-being. It's important to remember that these are generalizations, and there can be significant variation within both urban and rural contexts.

QUESTION BANK

Short Answer Questions

1. Define livelihood.
2. What are the major sources of livelihood in rural areas?
3. Name two primary sources of urban livelihood.
4. How does migration impact livelihood patterns?
5. What is the difference between formal and informal employment?
6. Mention one government scheme that supports rural livelihoods.
7. What role does agriculture play in rural livelihoods?
8. How does technology impact urban livelihoods?
9. What are the key components of sustainable livelihoods?
10. How do livelihood patterns differ between urban and rural areas?

Multiple Choice Questions (MCQs)

1. What does "livelihood" primarily refer to?

   a) Wealth accumulation b) A means of earning a living

   c) Government policies d) Education system

2. Which of the following is the main source of livelihood in rural areas?

   a) Agriculture b) IT services

   c) Factory work d) Banking

3. Which sector provides the majority of employment in urban areas?

   a) Manufacturing b) Agriculture

   c) Service sector d) Handicrafts

4. What is the main characteristic of informal employment?

   a) Job security

   b) Regular salary and benefits

   c) Lack of fixed working hours and security

   d) Government regulation

5. Which of the following is an example of a rural non-farm livelihood activity?
   a) Software development
   b) Tailoring
   c) Banking
   d) Import-export business
6. Which government program aims to improve rural livelihoods?
   a) Digital India
   b) MGNREGA
   c) Swachh Bharat Abhiyan
   d) Startup India
7. Which of the following is NOT a factor influencing urban livelihoods?
   a) Industrialization
   b) Climate change
   c) Infrastructure development
   d) Traditional farming
8. Which sector employs the largest number of people in rural India?
   a) Manufacturing
   b) IT sector
   c) Agriculture
   d) Tourism
9. What is the main challenge faced by people engaged in informal livelihoods?
   a) High salary
   b) Job security and lack of social benefits
   c) Government protection
   d) Fixed working hours
10. What is a major reason for rural-to-urban migration in India?
   a) Family traditions
   b) Entertainment opportunities
   c) Lack of employment in rural areas
   d) Availability of large farmlands

True/False Questions

1. Livelihood includes only employment and income. **(False)** *(It also includes assets, skills, and social resources)*
2. Agriculture is the sole livelihood option in rural areas. **(False)** *(There are also activities like animal husbandry, handicrafts, and rural industries)*
3. Urban livelihoods are mostly dependent on industries and the service sector. **(True)**

4. Informal employment in urban areas provides high job security. **(False)** *(It lacks job security and benefits)*
5. Sustainable livelihoods ensure long-term income generation. **(True)**
6. Migration from rural to urban areas can improve livelihood opportunities. **(True)**
7. Government policies do not influence livelihood patterns. **(False)** *(They play a crucial role in employment and skill development)*
8. Livelihood diversification helps reduce risks for rural households. **(True)**
9. The IT sector is a major source of employment in rural India. **(False)** (*It is more prominent in urban areas*)
10. The Mahatma Gandhi National Rural Employment Guarantee Act (MGNREGA) provides 100 days of wage employment in a year. **(True).**

4

Different Indicators to Study Livelihood Systems

Studying livelihood systems requires a multidimensional approach, as livelihoods are shaped by a variety of social, economic, environmental, and institutional factors. To understand how individuals and communities sustain their lives, researchers and policymakers use a range of indicators that help analyze both the opportunities available and the vulnerabilities faced by people. These indicators provide insights into the resources people use, the strategies they adopt, and the outcomes they achieve in their pursuit of a secure and dignified living. One of the foundational frameworks for analyzing livelihoods is the **Sustainable Livelihoods Framework (SLF)**, which identifies five core categories of capital or assets: **human capital**, **natural capital**, **financial capital**, **physical capital**, and **social capital**. Indicators related to each of these asset types are commonly used in livelihood studies.

Apart from these core assets, **livelihood outcome indicators** are also important. These include levels of food security, poverty status, employment stability, quality of life, savings and investment behavior, and ability to withstand shocks. These outcomes provide a picture of whether the livelihood strategies adopted are sustainable and equitable. **Institutional and policy indicators** are equally significant. These include access to government schemes, land rights, legal entitlements, effectiveness of rural development programs, and availability of extension services. The role of markets, financial institutions, and governance structures must also be considered when evaluating livelihood systems. In the context of **vulnerability**, indicators like frequency of crop failure, income variability, health risks, and exposure to natural disasters are used to assess how secure or fragile a livelihood is. Coping strategies, such as migration, asset liquidation, or diversification of income sources, are also analyzed.

Ultimately, livelihood indicators must be context-specific and sensitive to gender, caste, class, and geography. A well-designed livelihood study integrates both **quantitative data** (such as income levels, literacy rates, land size) and **qualitative insights** (such as perceptions of well-being, power dynamics, and cultural practices). By using a comprehensive set of indicators, researchers can gain a deeper understanding of livelihood systems and design more effective interventions to enhance resilience, reduce poverty, and promote inclusive development.

Livelihood systems are complex, and various indicators can be used to study them. These indicators can be broadly categorized into:

1. **Natural Capital:** Natural capital refers to the natural resource base that supports livelihoods, particularly in rural areas. Relevant indicators include access to arable land, quality and quantity of water, forest resources, soil fertility, biodiversity, and exposure to environmental risks like drought or floods.

These indicators assess the availability and quality of natural resources essential for livelihoods.

- **Land**
  - Land tenure security (e.g., percentage of households with formal land titles)
  - Land degradation (e.g., soil erosion rates, vegetation cover)
  - Land use patterns (e.g., agricultural land, forest cover)
  - Access to land (e.g., average landholding size)
- **Water**
  - Access to water resources (e.g., distance to water sources, percentage of households with access to potable water)
  - Water quality (e.g., levels of pollution)
  - Irrigation infrastructure (e.g., area under irrigation)
  - Water availability (e.g., rainfall patterns, groundwater levels)
- **Biodiversity**
  - Availability of wild foods and medicinal plants
  - Diversity of crops and livestock
  - Status of key ecosystem services (e.g., pollination, water purification)
- **Climate**
  - Rainfall patterns and variability
  - Temperature trends and extremes
  - Frequency and intensity of droughts, floods, and other extreme weather events

2. Physical Capital: Physical capital consists of infrastructure and goods that support livelihood activities. Indicators include access to roads, transportation, electricity, irrigation facilities, tools and machinery, housing quality, and communication technologies such as mobile phones or internet. These relate to the economic resources that people can use for consumption, investment, and coping with shocks. Indicators include household income levels, access to credit and savings, remittances, possession of livestock or other financial assets, and sources of cash income (e.g., farming, wage labor, small business). These indicators relate to the basic infrastructure and goods used to support livelihoods.

- **Infrastructure**
  - Access to roads and transportation networks (e.g., road density, travel time to markets)
  - Access to energy (e.g., percentage of households with electricity)
  - Access to communication technologies (e.g., mobile phone penetration, internet access)
  - Irrigation systems
  - Storage facilities
- **Technology**
  - Availability and use of agricultural equipment (e.g., tractors, plows)
  - Use of improved seeds and fertilizers
  - Adoption of sustainable farming practices
- **Housing**
  - Housing quality (e.g., type of construction materials, access to sanitation)
  - Overcrowding (e.g., number of people per room)

3. **Human Capital:** These reflect the education, skills, health, and labor capacity of individuals. Key indicators include literacy rates, level of educational attainment, vocational skills, access to healthcare, nutritional status, and household labor availability. Human capital is critical for accessing and maintaining productive employment and adapting to changing conditions. These indicators measure the skills, knowledge, health, and capacity for labour that individuals possess.

- **Education**
  - Literacy rates (e.g., percentage of adults who can read and write)
  - School enrolment rates (e.g., primary, secondary, and tertiary education)

  - Educational attainment (e.g., average years of schooling)
  - Skills training and vocational education
- **Health**
  - Life expectancy
  - Infant and child mortality rates
  - Nutritional status (e.g., prevalence of malnutrition, stunting, wasting)
  - Access to healthcare services (e.g., distance to health clinics, availability of doctors)
  - Prevalence of diseases (e.g., HIV/AIDS, malaria)
- **Labour**
  - Labour force participation rate
  - Unemployment rate
  - Skills and knowledge of the workforce

4. **Financial Capital:** These indicators assess the availability of financial resources to support livelihoods.

- **Income**
  - Average household income
  - Poverty rate (e.g., percentage of households below the poverty line)
  - Income inequality (e.g., Gini coefficient)
  - Sources of income (e.g., agriculture, wage labour, remittances)
- **Savings**
  - Percentage of households with savings accounts
  - Average savings levels
- **Credit**
  - Access to credit (e.g., percentage of households with loans)
  - Interest rates on loans
- **Remittances**
  - Inflow of remittances from migrant workers
- **Insurance**
  - Access to crop insurance or other forms of risk management

Social Capital: Social capital encompasses the networks, relationships, and social norms that help individuals access resources and cope with challenges. Indicators include participation in community groups or cooperatives, presence of self-help groups (SHGs), trust within communities, and support from family or neighbors during crises. These indicators measure the relationships, networks, norms, and trust that enable people to cooperate and achieve common goals.

- **Social Networks**
  - Membership in community groups and organizations (e.g., farmer cooperatives, women's groups)
  - Strength of social ties (e.g., frequency of social interactions, levels of trust)
  - Access to information and support through social networks
- **Governance**
  - Participation in decision-making processes
  - Transparency and accountability of local government
  - Rule of law and access to justice
- **Social Inclusion**
  - Equity in access to resources and opportunities (e.g., gender equality, inclusion of marginalized groups)
  - Levels of social cohesion and tolerance
- **Security**
  - Levels of crime and violence
  - Political stability
  - Absence of conflict

6. Livelihood Strategies: Indicators that describe the activities people undertake to make a living.

- **Agricultural practices:** Crop diversification, livestock management, use of technology.
- **Non-farm activities:** Engagement in wage labour, petty trade, artisanal production.
- **Migration:** Temporary or permanent migration for employment opportunities.
- **Resource management:** Sustainable use of natural resources.
- **Diversification:** The extent to which households engage in multiple livelihood activities.

7. Livelihood Outcomes: These indicators measure the results of livelihood strategies, reflecting improvements in well-being.

- **Increased income:** Improvement in household income levels.
- **Improved food security:** Reduced hunger and malnutrition. Indicators like Household Dietary Diversity Score (HDDS) or Food Consumption Score (FCS) can be used.
- **Improved health:** Reduced disease prevalence and improved access to healthcare.
- **Improved education:** Increased school enrolment and educational attainment.
- **Reduced vulnerability:** Increased resilience to shocks and stresses.
- **Sustainable resource management:** Improved environmental sustainability.

Considerations when Selecting Indicators

- **Context-specificity:** Indicators should be relevant to the specific livelihood systems being studied. A nomadic pastoralist system will require different indicators than an urban informal settlement.
- **Data availability:** Choose indicators for which data can be collected relatively easily and reliably.
- **Measurability:** Indicators should be quantifiable or at least amenable to qualitative assessment.
- **Sensitivity:** Indicators should be sensitive to changes in livelihood conditions.
- **Policy relevance:** Select indicators that can inform policy and program interventions aimed at improving livelihoods.

By using a combination of these indicators, researchers and practitioners can gain a comprehensive understanding of livelihood systems and identify opportunities for improvement. The specific indicators chosen will depend on the research question, the context, and the availability of data. It's important to consider how these indicators interact with each other, as livelihood systems are interconnected.

QUESTION BANK

Short Answer Questions

1. What are livelihood indicators?
2. Name two economic indicators used to study livelihood systems.

3. What is the importance of food security as a livelihood indicator?
4. How does access to education influence livelihood systems?
5. What role does infrastructure play in livelihood sustainability?
6. Mention one social indicator that affects livelihoods.
7. How is employment rate used to assess livelihood conditions?
8. Why is asset ownership an important livelihood indicator?
9. How does climate change impact livelihood systems?
10. What is the significance of income diversification in livelihoods?

Multiple Choice Questions (MCQs)

1. Which of the following is NOT a key indicator of livelihood systems?
   a) Income level
   b) Literacy rate
   c) Television ownership
   d) Employment rate
2. Which indicator measures the ability of people to afford basic needs?
   a) Economic growth rate
   b) Poverty level
   c) Road infrastructure
   d) Forest coverage
3. What does the term "livelihood security" primarily refer to?
   a) Having multiple bank accounts
   b) Stable access to food, income, and shelter
   c) Living in a city
   d) Owning luxury goods
4. Which of these is a **social** indicator of livelihoods?
   a) Agricultural productivity
   b) Household income
   c) Education level
   d) Employment rate
5. What is a common **environmental** indicator of livelihood sustainability?
   a) Soil fertility
   b) Inflation rate
   c) Population density
   d) Crime rate
6. How is food security measured as a livelihood indicator?
   a) By the number of supermarkets in a region
   b) By the availability, accessibility, and affordability of food
   c) By the number of restaurants in a city
   d) By food export statistics

7. Which economic factor can significantly affect livelihood systems?
   a) Unemployment rate
   b) Television viewership
   c) Internet penetration
   d) Number of political parties
8. Which sector is most commonly associated with rural livelihoods?
   a) Banking
   b) Agriculture
   c) IT sector
   d) Space research
9. How does infrastructure development impact livelihoods?
   a) It makes cities more expensive
   b) It improves access to markets, education, and healthcare
   c) It increases pollution levels
   d) It reduces government revenue
10. Why is skill development an important livelihood indicator?
   a) It helps people secure better employment opportunities
   b) It increases social media presence
   c) It ensures government subsidies
   d) It reduces the number of universities

True/False Questions

1. Income level is an important economic indicator of livelihoods. **(True)**
2. Climate change has no impact on livelihood systems. **(False)**
3. Education is considered a social indicator of livelihood. **(True)**
4. Infrastructure such as roads and electricity has no effect on livelihood sustainability. **(False)**
5. Food security is an essential factor in assessing livelihood conditions. **(True)**
6. Economic indicators include only cash income and bank balance. **(False)** *(They also include employment, asset ownership, and poverty levels.)*
7. A high unemployment rate indicates a weak livelihood system. **(True)**
8. Healthcare access does not affect livelihood security. **(False)**
9. Diversification of income sources helps improve livelihood resilience. **(True)**
10. Soil fertility is an environmental indicator affecting rural livelihoods. **(True)**

5

Agricultural Livelihood Systems (ALS) Meaning, Approach, Approaches and Framework

Agricultural Livelihood Systems (ALS) refer to the complex set of interactions between farming practices, economic activities, social structures, and environmental factors that influence the livelihoods of communities, particularly in rural areas. In regions like India, where a significant portion of the population depends on farming for sustenance and income, understanding ALS is crucial for planning sustainable rural development and poverty alleviation. The **meaning of Agricultural Livelihood Systems** extends beyond just crop cultivation. It includes a variety of interlinked activities such as livestock rearing, agroforestry, fisheries, horticulture, and non-farm support services like input supply, processing, and marketing. These systems are influenced by agro-ecological conditions, landholding patterns, socio-economic status, access to infrastructure, and institutional support. A household's livelihood strategy may be entirely agriculture-based or may combine farming with off-farm and non-farm income-generating activities.

The **approach to Agricultural Livelihood Systems** is holistic and people-centered. It emphasizes understanding the livelihood strategies of farming households in the context of the resources they possess, the constraints they face, and the opportunities available to them. Rather than focusing solely on increasing agricultural productivity, the ALS approach takes into account sustainability, equity, resilience, and diversification. These systems encompass a mix of agricultural and non-agricultural activities that contribute to the overall wellbeing of individuals and households.

Key Components of Agricultural Livelihood Systems

1. Agricultural Production:

- **Crops and Livestock**: The core of ALS involves the cultivation of crops and the raising of livestock. Varieties and practices may depend on climate, soil type, and market demand.

- **Sustainable Practices**: Increasing focus on sustainable farming techniques to enhance soil health, conserve water, and reduce chemical use.

2. Economic Activities

- **Income Generation**: In addition to farming, household members may engage in off-farm jobs, trade, or entrepreneurship to diversify income sources and improve financial stability.
- **Market Access**: Availability and access to markets are vital, as they influence the profitability of agricultural produce and other goods.

3. Social Structures

- **Community Networks**: Social capital, such as local organizations and cooperatives, can provide support, knowledge sharing, and improved access to resources.
- **Cultural Practices**: Traditional knowledge and cultural practices shape agricultural methods and can enhance community resilience.

4. Environmental Factors

- **Climate Change**: ALS must adapt to changing weather patterns, which can affect crop yields and livestock productivity.
- **Natural Resource Management**: Sustainable management of natural resources, such as water and soil, is crucial for the long-term viability of agricultural livelihoods.

5. Policy and Governance

- **Supportive Policies**: Government policies can impact access to resources, market opportunities, and infrastructure, all of which influence ALS.
- **Land Tenure Systems**: Secure land tenure is essential for encouraging investments in agriculture and improving livelihoods.

Challenges Facing Agricultural Livelihood Systems

- **Climate Change**: Increasing variability in weather patterns can lead to crop failures and food insecurity.
- **Economic Volatility**: Fluctuating market prices for both inputs and outputs can create instability for farmers.
- **Access to Resources**: Limited access to land, water, credit, and technology can hinder productivity and innovation.

- **Population Pressure**: Growing populations may increase the demand for food, putting additional stress on agricultural systems.
- **Social Inequities**: Marginalized groups, including women and indigenous communities, may face additional challenges in accessing resources and opportunities.

Agricultural Livelihood Systems play a vital role in sustaining rural communities and food security worldwide. Strengthening these systems requires a holistic approach that incorporates agricultural practices, economic diversification, social cohesion, and environmental stewardship. As challenges evolve, adaptive strategies and innovations will be crucial to enhancing resilience and sustainability within ALS.

Agricultural Livelihood Systems (ALS): Meaning and Approach

- **Meaning:** An Agricultural Livelihood System (ALS) encompasses all the activities, resources, and entitlements that enable a household to make a living through agriculture-related activities. It's broader than just farming and considers the interplay of various factors affecting a household's well-being.
    - It includes crop production, livestock rearing, fisheries, forestry, and related off-farm activities (e.g., agricultural processing, marketing, input supply).
    - It also takes into account the social, economic, and environmental context in which these activities take place. This includes access to markets, infrastructure, technology, information, credit, and natural resources (land, water).
    - Crucially, ALS considers the vulnerability context – the shocks, trends, and seasonality that affect livelihoods (e.g., climate change, market fluctuations, and policy changes).
- **Approach:** The ALS approach focuses on understanding how agricultural households:
    - **Access and utilize assets:** Land, labor, capital, social networks, and natural resources.
    - **Undertake activities:** Farming, livestock keeping, off-farm employment, etc.
    - **Are influenced by structures and processes:** Markets, policies, social norms, institutions.
    - **Achieve livelihood outcomes:** Increased income, improved food security, enhanced well-being, and sustainable resource management.

- **Approaches**
  - *Participatory Approach:* This involves working closely with farming communities to understand their needs, priorities, and constraints.
  - *Systems Thinking:* This recognizes the interconnectedness of different components within the ALS and how changes in one component can affect others.
  - *Livelihood Framework:* This provides a structured way to analyze ALS, considering assets, activities, vulnerability context, and outcomes.
- **Framework:** A typical ALS framework would include
  - **Asset Base:** Natural (land, water, biodiversity), Physical (infrastructure, equipment), Human (skills, knowledge, health), Financial (credit, savings), and Social (networks, institutions).
  - **Vulnerability Context:** Shocks (e.g., droughts, floods, pests), Trends (e.g., climate change, market fluctuations), Seasonality (e.g., rainfall patterns, crop cycles).
  - **Transforming Structures and Processes:** Institutions, policies, governance, culture, social norms.
  - **Livelihood Strategies:** Agricultural production, diversification, migration, off-farm employment.
  - **Livelihood Outcomes:** Increased income, improved food security, enhanced well-being, and sustainable resource use.

6

Definition of Farming Systems and Farming Based Livelihood Systems

What is a Farming System?

A farming system is a complex and interconnected set of components, activities, and interactions that occur on a farm. It's a holistic view of agriculture that considers not only the crops grown but also the livestock raised, the resources used, the environment, and the socio-economic factors that influence the farmer's decisions. It aims to optimize productivity and sustainability.

A farming system is a population of individual farm systems that have broadly similar resource endowments, enterprise patterns, household livelihoods, constraints, and for whom similar development strategies and interventions would be appropriate. It is a complex interwoven mesh of soil, plant, animal, people and other resources surrounding by a distinct boundary jointly managed by farm households.

A **farming system** refers to a complex, interrelated matrix of soil, water, crops, animals, labor, and other resources managed by farm households to increase productivity and ensure food security. It goes beyond crop production alone—it is a holistic concept that includes all on-farm and off-farm activities contributing to a family's livelihood. Farming-based livelihood systems emphasize the role of agriculture and allied sectors in supporting rural life and sustainability.

- It considers the interactions between different components of the farm (crops, livestock, soil, water, etc.) and how they are managed by the farm household.
- The aim is to optimize resource use, increase productivity, and improve the sustainability of agricultural production.

Farming-Based Livelihood System:

This is essentially an ALS where farming activities (crop production, livestock rearing, etc.) are the primary source of income and sustenance for the household. It highlights the centrality of farming within the broader livelihood strategy.

- In this system, the household's well-being is closely tied to the performance of the agricultural sector. Factors like access to inputs, markets, and technology directly impact their livelihood.
- Farming-based livelihood systems often include a mix of on-farm and off-farm activities to diversify income sources and reduce vulnerability.

A. Traditional Farming Systems

These systems generally rely on established techniques passed down through generations, with a strong connection to the local environment. They often prioritize sustainability and resourcefulness over maximizing yields in the short term.

- **Shifting Cultivation (Slash and Burn):**
  - *Description:* Farmers clear a plot of land (often by burning vegetation), cultivate it for a few years until soil fertility declines, and then abandon it to regenerate naturally. They then move to a new plot.
  - *Characteristics:* Low input (minimal fertilizer, pesticides), reliance on natural soil fertility, requires a large land area, and is sustainable only with long fallow periods.
  - *Where it's found:* Tropical rainforests, areas with low population density.
- **Subsistence Farming**
  - *Description:* Farmers grow enough food to feed themselves and their families, with little surplus for sale.
  - *Characteristics:* Small-scale, labour-intensive, uses simple tools, relies on family labour, and diverse crops are often grown for different needs.
  - *Where it's found:* Developing countries, rural areas.
- **Pastoral Nomadism**
  - *Description:* Herding livestock (cattle, sheep, goats, camels) and moving them from place to place in search of pasture and water.
  - *Characteristics:* Adaptation to arid and semi-arid environments, strong social structures related to livestock ownership, knowledge of grazing patterns.
  - *Where it's found:* Dry regions of Africa, Asia, and the Middle East.
- **Intensive Subsistence Farming**
  - *Description:* High labour input to maximize yields on small landholdings.

  - *With Wet Rice Dominant:* Rice is the main crop, often grown in flooded paddies.
  - *Without Wet Rice Dominant:* Other crops like wheat, barley, corn, or vegetables are grown.
- *Characteristics:* Careful land preparation, irrigation (where applicable), use of animal manure, crop rotation.
- *Where it's found:* Densely populated regions of Asia, such as India, China, and Southeast Asia.

B. Modern Farming Systems

Modern Farming Systems refer to advanced agricultural practices and technologies that aim to maximize productivity, efficiency, and sustainability in farming. These systems integrate scientific knowledge, mechanization, precision tools, and environmentally sound practices to meet the growing global demand for food, fiber, and fuel. These systems emphasize maximizing productivity, efficiency, and profitability through technology, scientific advancements, and economies of scale.

Definition of Modern Farming Systems

Modern farming systems are technologically driven agricultural systems that utilize innovations such as improved seeds, chemical inputs (fertilizers and pesticides), irrigation methods, mechanization, and data-based decision-making tools to increase farm productivity while optimizing resource use and minimizing environmental impact.

Key Features of Modern Farming Systems

1. **Use of High-Yielding Varieties (HYVs):** Improved crop varieties developed through plant breeding to produce more grains, resist diseases, and adapt to specific environments.
2. **Mechanization:**Use of modern machinery like tractors, combine harvesters, seed drills, and sprayers to reduce manual labor and improve efficiency.
3. **Precision Agriculture:**Application of technologies such as GPS, drones, remote sensing, and sensors to monitor and manage crops and soils with high precision.
4. **Efficient Irrigation Systems:**Techniques like drip irrigation, sprinkler systems, and automated irrigation scheduling that conserve water and deliver it directly to plant roots.
5. **Integrated Nutrient Management (INM):** Combining organic and inorganic fertilizers for balanced soil fertility and crop nutrition.

6. **Integrated Pest Management (IPM):** Use of biological control, cultural practices, and selective pesticides to manage pest populations economically and with minimal harm to the environment.
7. **Greenhouses and Controlled Environment Agriculture (CEA):** Growing crops in protected environments with controlled temperature, humidity, and light to extend growing seasons and increase yield.
8. **Sustainable and Climate-Smart Practices:** Techniques that improve resilience to climate change, reduce greenhouse gas emissions, and promote soil and water conservation.
9. **Use of Biotechnology and Genetic Engineering:** Application of modern biology to develop genetically modified (GM) crops with desirable traits such as pest resistance or drought tolerance.
10. **Farm Management Software and ICT Tools:** Digital tools and mobile apps that help farmers in record-keeping, decision-making, weather forecasting, and market access.

Advantages of Modern Farming Systems

- Increased crop yield and farm profitability
- Reduced labor and time requirements
- Improved resource-use efficiency (water, nutrients, energy)
- Enhanced pest and disease control
- Better market access and traceability
- Capacity to farm larger areas with fewer resources

Challenges and Considerations

- High initial investment and maintenance cost
- Risk of environmental degradation due to overuse of chemicals
- Dependency on non-renewable inputs
- Knowledge and skill requirements for technology adoption
- Socio-economic inequalities among smallholder and large-scale farmers

C. Intensive Commercial Farming

- *Description:* Large-scale production of crops or livestock for sale in the market.
- *Characteristics:* High use of machinery, fertilizers, pesticides, irrigation, and often genetically modified (GM) crops. Focus on monoculture (growing a single crop). Aims to maximize yield and profit.
- *Types*

Grain Farming: Wheat, corn, soybeans, etc. (e.g., in the US Midwest, Canadian Prairies)

Livestock Ranching: Cattle, sheep, etc. (e.g., in the Western US, Argentina)

Dairy Farming: Milk production (e.g., in Europe, North America)

Fruit and Vegetable Farming: (e.g. California, Mediterranean)

- *Where it's found:* Developed countries, areas with favourable climate and soil conditions.

1. Plantation Agriculture

- *Description:* Large-scale farming of cash crops for export, often in tropical regions. Historically associated with colonial exploitation.
- *Characteristics:* Monoculture, reliance on hired labour, significant capital investment, focuses on export markets.
- *Crops:* Rubber, tea, coffee, cocoa, bananas, sugarcane, palm oil.
- *Where it's found:* Tropical regions of Asia, Africa, and Latin America.

2. Factory Farming (Intensive Animal Production)

- *Description:* Raising large numbers of livestock in confined spaces, with a focus on rapid growth and efficiency.
- *Characteristics:* High density of animals, use of specialized feed, antibiotics, and hormones, and often causes environmental concerns (manure management, pollution).
- *Animals:* Chickens, pigs, cattle.
- *Where it's found:* Developed countries, increasingly in developing countries.

3. Precision Agriculture

- *Description:* Using technology (GPS, sensors, drones, data analytics) to optimize inputs (fertilizer, water, pesticides) and management practices based on site-specific conditions within a field.
- *Characteristics:* Increased efficiency reduced environmental impact, data-driven decision-making.
- *Where it's found:* Developed countries, large-scale farms.

4. Vertical Farming

- *Description:* Growing crops in vertically stacked layers, often indoors in a controlled environment.

- *Characteristics:* High-tech, uses artificial lighting, hydroponics or geoponics (growing plants without soil), year-round production, reduced water use, and can be located in urban areas.
- *Where it's found:* Urban areas, research facilities, and areas with limited arable land.

Emerging Trends (Hybrids and New Approaches)

- **Agroforestry:** Integrating trees and shrubs into agricultural systems.
- **Conservation Agriculture:** Minimizing soil disturbance, maintaining soil cover, and diversifying crop rotations.
- **Organic Farming:** Avoiding synthetic fertilizers, pesticides, and GMOs, focusing on soil health and biodiversity. While it can be considered both a traditional-inspired approach and a modern market-driven system.
- **Permaculture:** Designing sustainable human settlements & agricultural systems modelled on natural ecosystems.

Key Differences Summarized

Feature	Traditional Farming	Modern Farming
Scale	Small to Medium	Large to Very Large
Labour	Labour-intensive	Capital-intensive
Technology	Simple tools, traditional knowledge	Advanced machinery, technology
Inputs	Low external inputs	High external inputs
Crop Diversity	High	Often low (monoculture)
Sustainability	Potentially sustainable	Can be unsustainable if not managed well
Primary Goal	Subsistence or local markets	Commercial sale, profit maximization

II. Key Components of a Farming System/ Farming-Based Livelihood System

Farming System/ Farming-Based Livelihood System can broadly categorize the components into:

- **Natural Resources:** These are the foundation of the system.
- **Crop Production:** The cultivation of various crops.
- **Livestock Production:** Raising animals for various products and services.
- **Other Enterprises:** Activities that supplement income.
- **Human Component:** The farmer and their family/labour.
- **Management Practices:** How the farmer makes decisions and manages resources.
- **External Factors:** Market forces, policies, and climate.

A farming system refers to the dynamic combination of natural, economic, and human resources used by farmers to produce crops and raise livestock, thereby securing their food, income, and livelihood. Farming-based livelihood systems are broader in scope, encompassing both agricultural and non-agricultural activities that sustain rural households. The farming system is highly diverse and location-specific, influenced by environmental conditions, socio-economic factors, and available technologies. Understanding its key components is essential for developing sustainable, resilient, and efficient agricultural practices.

One of the fundamental components of a farming system is **natural resources**, which include land, water, soil, climate, and biodiversity. These resources form the ecological foundation upon which farming activities depend. Land provides the physical space for crop cultivation and livestock rearing, while water is essential for irrigation and animal care. Soil health determines crop productivity and can be enhanced through organic amendments and conservation practices. Climate and weather patterns influence cropping calendars, pest incidence, and overall farm productivity. Biodiversity, including crop varieties, livestock breeds, and beneficial organisms, plays a key role in maintaining ecological balance and long-term farm sustainability.

Human capital is another crucial component, encompassing the labor, knowledge, skills, and experience of the farm family and hired workers. The availability of labor affects the scale and intensity of agricultural operations. Farmers' indigenous knowledge and technical skills guide the selection of crops, livestock management, and the use of appropriate technologies. Capacity-building through training and extension services can significantly enhance productivity and the adoption of sustainable practices.

Crop production systems are central to most farming operations. These include the types of crops grown, cropping patterns such as monocropping, intercropping, and crop rotation, and crop management techniques like fertilization, irrigation, weed control, and harvesting. Crop choices are determined by agro-climatic suitability, household food needs, and market demand. The efficiency and sustainability of crop production are improved through soil fertility management, integrated pest management, and the use of improved seed varieties.

In many farming systems, **livestock** plays a complementary role. Animals such as cattle, goats, poultry, and pigs provide milk, meat, eggs, draught power, manure, and additional income. Livestock also serve as a financial reserve during times of distress and contribute to diversified livelihoods. Proper feeding, health care, and breeding management are essential to ensure the productivity and profitability of livestock enterprises.

Integrated Farming Systems (IFS) represent a holistic approach where various components such as crops, livestock, horticulture, fisheries, agroforestry, and other enterprises are combined on a single farm. This integration maximizes resource use efficiency, enhances income, and reduces risk by diversifying income sources. For example, combining rice cultivation with fish farming or integrating fruit trees with vegetables and poultry creates a more sustainable and resilient farming model.

The availability and accessibility of **capital and inputs** also significantly influence the performance of farming systems. These include financial capital (such as savings, credit, and insurance), physical inputs (such as seeds, fertilizers, pesticides, tools, and machinery), and infrastructure (such as irrigation facilities, storage structures, and rural roads). Access to timely and affordable inputs and infrastructure is vital for improving farm productivity and reducing post-harvest losses.

Market access and value chains are essential for transforming agricultural production into a profitable livelihood. Farmers need reliable access to input markets, output markets, and fair pricing mechanisms. Strengthening value chains through processing, grading, packaging, and marketing helps farmers capture higher value and reduces dependence on middlemen. Access to market information, digital platforms, and cooperatives further enhances marketing efficiency and bargaining power.

Supportive **institutional and policy frameworks** also shape farming systems. Government programs, subsidies, minimum support prices, crop insurance schemes, and land tenure security play a crucial role in incentivizing sustainable practices and ensuring farmers' welfare. Agricultural extension services, research institutions, and non-governmental organizations contribute to capacity building, technology dissemination, and community mobilization.

Social and cultural aspects form another layer of the farming system. **Socio-economic factors** such as household size, education level, gender roles, and traditional norms influence decision-making, resource allocation, and adoption of innovations. Social networks, community-based organizations, and farmer cooperatives provide mutual support, risk sharing, and collective bargaining strength, enhancing the overall resilience of the farming system.

Environmental sustainability must be integrated into all farming system components. Practices such as conservation agriculture, organic farming, agroforestry, water harvesting, and the use of renewable energy contribute to sustainable resource management. Climate-smart agriculture, which promotes adaptation and mitigation strategies, is increasingly vital in the face of climate variability and environmental degradation.

A farming system is an interconnected web of biophysical, human, economic, and institutional components that interact to produce food and support rural livelihoods. A deep understanding of these components is essential for designing interventions that enhance productivity, sustainability, and resilience in the agricultural sector. Strengthening each component in a coordinated manner can lead to improved food security, income stability, and overall rural development.

QUESTION BANK

Short Answer Questions

1. Define Agricultural Livelihood Systems (ALS).
2. What are the key components of an Agricultural Livelihood System?
3. What is a farming system?
4. List any four components of a farming-based livelihood system.
5. How does livestock contribute to a farming system?
6. What is the significance of integrated farming systems (IFS)?
7. Mention two natural resources critical to farming systems.
8. How does ALS contribute to rural development?
9. Name two major approaches used in studying ALS.
10. What role does sustainability play in Agricultural Livelihood Systems?
11. How does climate change impact ALS?
12. What is the significance of the Sustainable Livelihood Approach (SLA) in ALS?
13. Mention one challenge faced by agricultural livelihood systems.
14. How do government policies influence ALS?
15. What is the role of diversification in Agricultural Livelihood Systems?

Multiple Choice Questions (MCQs)

1. What does Agricultural Livelihood Systems (ALS) primarily focus on?
    a) Urban employment
    b) Industrial growth
    c) Sustainable farming and rural livelihoods
    d) Space research

2. Integrated Farming System (IFS) aims to
   a) Focus only on monoculture b) Reduce farm diversification
   c) Maximize resource use efficiency d) Eliminate animal husbandry
3. Livestock in farming systems primarily provides
   a) Pesticides b) Organic manure
   c) Electricity d) Subsidies
4. Which of the following is an example of a cropping pattern?
   a) Crop harvesting b) Crop marketing
   c) Intercropping d) Soil testing
5. Farm mechanization leads to:
   a) Reduced productivity b) Increased labor costs
   c) Reduced time and effort d) Soil erosion
6. A major benefit of market linkage in farming is:
   a) Decrease in profit b) Limited production
   c) Better price realization d) Reduced irrigation
7. Which of the following is NOT part of natural resources in farming?
   a) Air b) Soil
   c) Biodiversity d) Mobile phones
8. Agroforestry is a component of:
   a) Traditional farming only b) Animal husbandry
   c) Integrated Farming Systems d) Irrigation systems
9. The term 'sustainable farming' refers to farming that
   a) Depends only on chemical inputs
   b) Aims to improve environmental health
   c) Focuses on a single crop
   d) Uses non-renewable resources
10. Farming-based livelihood systems mainly support which group in India?
   a) Urban business owners b) Industrial workers
   c) Rural households d) Software professionals

11. Which of the following is NOT a component of ALS?

    a) Soil fertility
    b) Internet usage
    c) Market access
    d) Agricultural labor

12. The Sustainable Livelihood Approach (SLA) in ALS focuses on:

    a) Short-term economic growth
    b) Reducing rural poverty and improving resilience
    c) Increasing government tax revenue
    d) Urban infrastructure development

13. Which approach is commonly used to analyze Agricultural Livelihood Systems?

    a) Linear approach
    b) Systems approach
    c) Random approach
    d) Hypothetical approach

14. What is a key factor in ensuring a sustainable Agricultural Livelihood System?

    a) Dependence on a single crop
    b) Diversification of income sources
    c) Reducing the workforce in farming
    d) Importing all agricultural products

15. Which framework is commonly used in Agricultural Livelihood Systems?

    a) Economic Growth Model
    b) Sustainable Livelihood Framework (SLF)
    c) Social Media Marketing Model
    d) Political Development Model

16. The ALS approach integrates which key aspect?

    a) Natural resources, human resources, and markets
    b) Only government policies
    c) Urban business models
    d) Digital marketing strategies

17. What is one major challenge to Agricultural Livelihood Systems?
    a) Stable government support
    b) Climate change and water scarcity
    c) Unlimited access to credit
    d) Low population in rural areas
18. How does ALS contribute to economic growth?
    a) By increasing urban unemployment
    b) By reducing food production
    c) By generating employment in rural areas
    d) By promoting industrial pollution
19. What is the main goal of ALS frameworks?
    a) To maximize rural poverty
    b) To promote sustainable agricultural practices and income stability
    c) To discourage farming activities
    d) To eliminate government support for farmers
20. **Which of the following is NOT a component of a farming system?**
    a) Soil
    b) Internet
    c) Livestock
    d) Climate

True/False Questions

1. Agricultural Livelihood Systems focus only on crop production. **(False)** *(They include livestock, fisheries, and other rural activities.)*
2. Climate change has no impact on ALS. **(False)**
3. The Sustainable Livelihood Framework (SLF) is widely used in ALS. **(True)**
4. Diversification of agricultural income reduces risks for farmers. **(True)**
5. Market access plays no role in ALS development. **(False)**
6. The Systems Approach is often used to analyze ALS. **(True)**
7. ALS frameworks focus on increasing agricultural sustainability. **(True)**
8. Natural resources are a critical component of Agricultural Livelihood Systems. **(True)**
9. Government policies have no influence on ALS. **(False)**
10. The Sustainable Livelihood Approach (SLA) is irrelevant to agricultural development. **(False)**

7

Prevalent Farming Systems in India Contributing to Livelihood

Prevalent Farming Systems in India Contributing to Livelihood

India has a diverse range of farming systems, shaped by agro-climatic zones, socio-economic conditions, and cultural practices. India, with its diverse agro-climatic zones and socio-economic conditions, supports a variety of **farming systems** that play a crucial role in providing **food, income, and employment** to a large segment of the rural population. These systems vary by region, resource availability, landholding size, and climatic conditions, and are vital in ensuring **sustainable livelihoods** for millions of small and marginal farmers. Here are some prevalent examples:

- **Rice-Wheat System:** Predominant in the Indo-Gangetic Plains. It involves rotating rice in the wet season and wheat in the dry season. This system is crucial for food security but faces challenges related to water depletion and soil degradation.
- **Rainfed Farming Systems:** Found in arid and semi-arid regions. Farmers rely on rainfall for crop production. Common crops include millets, pulses, and oilseeds. Drought-resistant varieties and water conservation techniques are essential.
- **Mixed Farming Systems:** Integrate crop production with livestock rearing. This system is common across India and provides multiple income streams. Livestock provides manure for soil fertility and draught power, while crops provide feed for animals.
- **Horticultural Systems:** Focus on the production of fruits, vegetables, and flowers. These systems are often found in regions with favorable climate and access to markets. They can generate high incomes but require significant investment and management skills.
- **Plantation Systems:** Involve the cultivation of cash crops like tea, coffee, rubber, and spices. These systems are concentrated in specific regions with suitable agro-climatic conditions.

- **Coastal Farming Systems:** Involve agriculture, aquaculture, and fisheries. Farmers often cultivate rice, coconuts, and vegetables, while also engaging in fish farming or fishing.
- **Agroforestry Systems:** Integrate trees into agricultural landscapes. Trees can provide shade, improve soil fertility, and generate additional income from timber, fruits, or other products.
- **Integrated Farming Systems (IFS):** These combine two or more farming systems, like agriculture, animal husbandry, fisheries, forestry, etc., to enhance resource use efficiency and sustainability. For example, integrating fish farming with rice cultivation or poultry farming with horticulture.
- These systems contribute significantly to the livelihoods of millions of farmers and agricultural laborers in India. However, they face numerous challenges, including:
- **Climate Change:** Increased frequency and intensity of droughts, floods, and heat waves.
- **Water Scarcity:** Depletion of groundwater resources and increasing competition for water.
- **Soil Degradation:** Erosion, nutrient depletion, and salinization.
- **Market Volatility:** Fluctuations in crop prices and lack of access to reliable markets.
- **Small Landholdings:** Limited economies of scale and difficulty in adopting new technologies.
- **Lack of Access to Credit and Insurance:** Increased vulnerability to shocks and limited investment capacity.

Farming Systems in India Contributing to Livelihood

1. Crop-Based Farming System

The crop-based farming system is one of the most predominant and traditional forms of agriculture in India. It primarily revolves around the cultivation of crops as the main agricultural activity, and is widely practiced across various agro-climatic regions of the country. This system is characterized by its focus on growing cereals, pulses, oilseeds, vegetables, fruits, and cash crops depending on the geographical, climatic, and socio-economic conditions of the area. In India, the crop-based farming system is particularly dominant in the fertile plains of the Indo-Gangetic region, including states like Punjab, Haryana, Uttar Pradesh, Bihar, and West Bengal. These areas benefit from rich alluvial soils, favorable rainfall patterns, and well-established irrigation infrastructure, making them highly suitable for intensive crop cultivation. Rice and wheat are the most widely grown crops in these regions, often in a rotation system that supports soil fertility and maximizes land use efficiency.

The diversity of India's agro-ecological zones allows for a wide range of cropping patterns. In central and western India, crops such as maize, jowar (sorghum), bajra (pearl millet), soybean, and cotton are commonly cultivated, especially in rainfed conditions. In the southern and coastal states, crops like paddy, sugarcane, groundnut, ragi, and coconut dominate. Meanwhile, northeastern states and hill regions grow niche crops such as buckwheat, millets, and horticultural produce adapted to the local terrain and climate. Crop-based systems in India can be either monoculture or multiple cropping systems. Monoculture, or the cultivation of a single crop season after season, is commonly seen in commercial agriculture, such as sugarcane in Maharashtra or cotton in Gujarat. However, this can sometimes lead to problems like nutrient depletion and pest build-up. In contrast, multiple cropping systems, such as intercropping, mixed cropping, and crop rotation, are more sustainable and help farmers reduce risk and increase income. Intercropping systems such as maize + legumes or wheat + mustard are popular in several regions for efficient land and resource use.

In many parts of India, the crop-based farming system is subsistence-oriented, especially among small and marginal farmers who grow food primarily for household consumption. However, with improved market access, infrastructure, and government support, commercial crop cultivation has gained momentum. Cash crops like sugarcane, cotton, tea, coffee, and spices are now major contributors to farm income in several states. The contribution of crop-based farming to rural livelihoods is significant. It ensures food security, generates employment in rural areas, and provides raw materials for agro-industries. Seasonal labor requirements during sowing, weeding, harvesting, and post-harvest handling create income opportunities for landless laborers and women. Additionally, surplus produce is often sold in local markets or procurement centers, offering monetary benefits to farming households.

Despite its importance, the crop-based farming system in India faces several challenges. These include dependence on monsoon rains, declining soil fertility, overuse of chemical fertilizers and pesticides, pest outbreaks, and fluctuating market prices. To address these issues, the government and research institutions promote sustainable practices such as organic farming, integrated nutrient and pest management, conservation agriculture, and the adoption of high-yielding and climate-resilient crop varieties. The crop-based farming system remains a cornerstone of Indian agriculture. It supports millions of rural livelihoods, contributes to national food security, and forms the foundation for integrated and diversified farming approaches. Strengthening this system through better technology, infrastructure, and policy support is essential for achieving inclusive and sustainable agricultural development in India.

This is the most dominant farming system in India, especially in fertile plains such as the Indo-Gangetic region, Punjab, Haryana, and parts of West Bengal.

- **Main features:** Cultivation of cereals (rice, wheat, maize), pulses, oilseeds, and cash crops (sugarcane, cotton).
- **Contribution to livelihood:** Provides staple food, income from surplus sale, and employment in planting, weeding, and harvesting.

2. Crop-Livestock Farming System

The crop-livestock farming system is one of the most traditional and widely adopted mixed farming systems in India. It involves the integration of crop cultivation and livestock rearing on the same farm, offering a balanced and mutually supportive livelihood strategy for millions of rural households. This system is particularly significant in semi-arid, arid, and rainfed regions where dependence on a single enterprise may not ensure economic and food security due to climatic uncertainties and limited resources. In the crop-livestock system, both components are interdependent and complementary. Crops provide fodder and feed for the animals, while livestock contribute manure for soil fertility, draught power for field operations, and additional income from products like milk, meat, eggs, wool, and hides. This integration enhances resource recycling, increases overall farm productivity, and reduces risk, making it a sustainable model for small and marginal farmers who dominate the Indian agricultural landscape.

This system is extensively practiced across various states in India. In Rajasthan, Gujarat, and Madhya Pradesh, for instance, cereals such as jowar, bajra, and maize are cultivated along with the rearing of cattle, goats, and sheep. In the southern states like Tamil Nadu and Karnataka, farmers combine paddy or groundnut cultivation with dairy farming. In eastern India, especially in Bihar, Jharkhand, and Odisha, mixed farms often include paddy fields and livestock like buffaloes and backyard poultry. The crop-livestock system plays a crucial role in ensuring year-round income and nutritional security. While crops are harvested seasonally, livestock can provide continuous returns through daily milk production or periodic sale of animals. Manure from livestock helps maintain soil health and reduces the dependence on costly chemical fertilizers. In traditional systems, animal power is still used for ploughing and transporting goods, particularly in less mechanized regions.

Another major advantage of this system is employment generation. It provides work opportunities for family members throughout the year, including women and elderly members, who often take care of livestock management. This inclusive nature of labor distribution within the household makes it an important livelihood strategy in rural India. However, the crop-livestock farming system also faces challenges. Shrinking grazing lands, fodder shortages, low animal productivity, poor veterinary services, and limited access to markets and credit facilities hinder its potential. Climate variability, especially droughts and floods, impacts both crops and livestock simultaneously, increasing vulnerability among dependent households.

To address these issues, integrated approaches are being promoted through government schemes and agricultural extension services. Programs like the Rashtriya Krishi Vikas Yojana (RKVY), National Livestock Mission (NLM), and various state-level dairy development initiatives aim to strengthen the crop-livestock interface by improving livestock breeds, enhancing fodder availability, and linking farmers to markets. In recent years, scientific interventions such as silage making, fodder banks, rotational grazing, composting, and improved housing for animals have been encouraged. These innovations help in optimizing livestock production without compromising the crop component. Moreover, agri-startups and cooperatives have played a role in value addition, such as producing ghee, paneer, and organic manure, which fetch better prices for farmers.

Widely practiced across India, especially in rainfed and semi-arid regions such as Rajasthan, Gujarat, and Madhya Pradesh.

- **Main features:** Integration of crops with cattle, goats, poultry, or sheep.
- **Livelihood impact:** Livestock provides milk, meat, manure, and draught power, while crops provide fodder—ensuring year-round income and food security.

3. Crop-Fish Farming System

The **crop-fish farming system** is a form of integrated agriculture in which crop cultivation is combined with fish culture either concurrently or sequentially. It is particularly well-suited for regions with abundant water resources, high rainfall, or low-lying fields that retain water for long periods. This system plays a crucial role in enhancing **farm income, resource use efficiency, and nutritional security**, especially for small and marginal farmers in rural India. In this system, rice cultivation is most commonly integrated with fish farming, particularly in waterlogged paddy fields. Known as **rice-fish farming**, this method has been traditionally practiced in parts of eastern India such as **West Bengal, Assam, Odisha, and Chhattisgarh**, where natural water availability allows for the coexistence of fish and crops. Other combinations include fish farming in pond dykes adjacent to vegetable or fruit crops, and seasonal fish culture in rainfed lowlands after the crop harvest. The basic principle of crop-fish integration lies in the **synergistic relationship** between the two components. Crops benefit from the nutrient-rich water enhanced by fish excreta, while fish gain from the shelter and food sources provided by the crop environment. For example, in paddy-fish systems, fish feed on weeds, insect larvae, and crop residues, helping in natural pest control. Their movement also aerates the water and improves soil structure, indirectly aiding in better crop yields.

A typical crop-fish farming setup involves creating trenches or ponds within or adjacent to crop fields, ensuring water retention for fish culture. Common fish species used include **Rohu, Catla, Mrigal**, and fast-growing exotic species like **Tilapia and Common Carp**. These are often stocked during the crop-growing season and harvested along with or after the crops. This integrated system has several advantages. It allows **diversification of income**, reduces dependence on a single source of livelihood, and **enhances protein intake** for rural households through fish consumption. Moreover, it contributes to the efficient use of on-farm resources like water, labor, and space. Women and elderly family members often participate in fish feeding and maintenance, contributing to **gender-inclusive farming practices**. Despite its benefits, crop-fish farming in India faces several challenges. These include inadequate technical knowledge among farmers, poor water management, predation of fish in open systems, and lack of access to quality fingerlings and markets. Water pollution and pesticide use in crops can also adversely affect fish health and productivity.

To promote this farming system, several initiatives have been taken by the government and agricultural universities. The **National Fisheries Development Board (NFDB)**, under the **Blue Revolution scheme**, supports integrated farming systems including aquaculture. State agricultural universities and Krishi Vigyan Kendras (KVKs) offer training on pond construction, fish stocking, and integrated nutrient management. In recent years, **climate-smart models** of crop-fish farming are being developed to ensure sustainability and resilience against climate risks such as floods and droughts. Farmer producer organizations (FPOs) and cooperatives have also helped in improving fish marketing and value addition through dried fish, fish pickle, and processed products.

Practiced in water-abundant regions like West Bengal, Assam, Odisha, and Kerala.

- **Main features:** Paddy fields are combined with fish cultivation in ponds or trenches.
- **Livelihood benefit:** Dual income source from fish and crops, with minimal additional inputs; fish improves protein availability.

4. Crop-Horticulture System

The **crop-horticulture farming system** is an increasingly popular form of integrated agriculture in India, combining the cultivation of food or cash crops with horticultural crops such as fruits, vegetables, flowers, medicinal plants, and spices. This system is highly suitable for regions with diverse agro-climatic conditions and is particularly beneficial for small and marginal farmers aiming to enhance income, nutrition, and employment from limited landholdings. In this system, horticultural crops are introduced alongside or in rotation with traditional field crops such as cereals (rice, wheat, maize), pulses, or oilseeds. The primary objective is to **diversify farm output**, optimize land and resource utilization, and generate **higher returns per unit area** compared to monoculture systems. This integration not only stabilizes farm income but also supports **nutritional security** by increasing the availability of fruits and vegetables in rural diets.

The crop-horticulture system is widely practiced in several parts of India. In the **hilly regions** of Himachal Pradesh, Uttarakhand, and Jammu & Kashmir, farmers grow apples, plums, pears, and vegetables alongside maize, wheat, or barley. In **western India**, states like Maharashtra and Gujarat combine cotton, groundnut, or sorghum with mangoes, bananas, pomegranates, and vegetables like onions and tomatoes. In **southern India**, especially in Tamil Nadu and Karnataka, paddy and sugarcane are interspersed with coconut, areca nut, spices, and vegetables. The **North-Eastern states**, rich in biodiversity, grow pineapple, ginger, turmeric, and oranges along with rice and millets. One of the major advantages of this system is its **economic viability**. Horticultural crops often fetch **higher market prices**, have **shorter harvesting cycles**, and provide multiple harvests in a year. Vegetables and floriculture, in particular, offer **quick returns**, enabling farmers to maintain a steady cash flow. Furthermore, the system is **labor-intensive**, generating significant rural employment, particularly for women and youth.

Another key benefit of the crop-horticulture system is **soil health improvement**. Many horticultural crops, especially legumes and cover crops, enrich the soil with organic matter and help prevent erosion, especially on sloped lands. The cultivation of perennial fruit trees and plantation crops also contributes to **carbon sequestration** and **climate resilience**. However, the adoption of this system is not without challenges. Farmers often face constraints such as

lack of technical knowledge, **limited access to quality planting material**, **inadequate post-harvest infrastructure**, and **price volatility** in horticultural markets. Perishability of horticultural produce also necessitates timely harvesting, storage, and marketing, which many rural areas are ill-equipped for. To promote crop-horticulture integration, the Indian government has launched several initiatives, including the **National Horticulture Mission (NHM)**, **MIDH (Mission for Integrated Development of Horticulture)**, and **Rashtriya Krishi Vikas Yojana (RKVY)**. These programs aim to provide farmers with financial support, training, improved seeds and planting material, and post-harvest management facilities. Extension services, agri-startups, and Farmer Producer Organizations (FPOs) have also played a key role in spreading awareness and connecting growers to markets.

Increasingly popular in states like Maharashtra, Tamil Nadu, Himachal Pradesh, and Jammu & Kashmir.

- **Main features:** Integration of fruits (mango, banana, guava), vegetables (tomato, brinjal), and plantation crops (coconut, areca nut) with food crops.
- **Livelihood impact:** Higher returns per unit area, employment in nursery raising and value addition, especially beneficial for small farmers.

5. Crop-Livestock-Poultry System

The **Crop-Livestock-Poultry Farming System** is a holistic and integrated farming approach that combines three major agricultural components—**crop cultivation, livestock rearing, and poultry farming**—on a single farm. This diversified system is designed to optimize the use of available resources, reduce dependence on external inputs, and increase the overall resilience and profitability of smallholder farming households in India. This system is particularly beneficial for **small and marginal farmers**, who often operate under resource-constrained conditions. By integrating crops with animals and poultry, farmers are able to achieve **multiple streams of income**, better food security, and improved nutrient cycling on the farm. Each component supports the other: crops provide feed and fodder, livestock contribute manure and traction, while poultry offer a continuous supply of eggs, meat, and organic inputs for soil enrichment.

In India, this system is practiced across **diverse agro-climatic zones**. For example:

- In **eastern India**, especially West Bengal, Odisha, and Bihar, rice-based cropping systems are integrated with cattle and backyard poultry such as local hens or ducks.

- In **southern states** like Tamil Nadu and Andhra Pradesh, paddy and pulses are combined with dairy animals and poultry units (broilers or layers).
- In **central India**, crops like millets and pulses are grown alongside goats, cattle, and country chickens, providing robust livelihood support in rainfed areas.

One of the significant advantages of this system is its **economic stability**. While crop income is seasonal, livestock and poultry offer more **frequent and predictable returns**. Milk and eggs provide daily or weekly income, while animals and poultry can be sold during financial emergencies. This reduces the risks of crop failure due to droughts or floods and **buffers farm families** against market or climate shocks. The system also offers **ecological benefits**. Animal dung and poultry droppings enrich the soil, reducing the need for chemical fertilizers. Crop residues can be used as fodder, while surplus grains serve as poultry feed. This internal recycling of nutrients minimizes waste and improves **resource-use efficiency**. Moreover, the diversified nature of the system encourages better **labour utilization**, where different family members can participate in various activities year-round, promoting inclusive employment.

Women, in particular, play a significant role in managing backyard poultry and dairy animals. This not only enhances their contribution to household income but also **empowers women economically** and socially in rural settings. Despite its benefits, the crop-livestock-poultry system faces several constraints. These include **limited access to veterinary care**, inadequate fodder and feed supply, **lack of credit and insurance**, and **market fluctuations** for livestock and poultry products. Diseases, especially among poultry, can spread rapidly and affect the sustainability of the system. Also, small farmers may struggle with space and infrastructure for managing livestock and poultry hygienically.

To address these challenges, government schemes such as the **National Livestock Mission**, **Dairy Entrepreneurship Development Scheme**, **Rashtriya Gokul Mission**, and **backyard poultry development programs** have been implemented to provide financial assistance, training, and technical support. Research institutions and Krishi Vigyan Kendras (KVKs) also play a key role in demonstrating improved breeds, vaccination schedules, and balanced nutrition practices. In recent years, the adoption of **Integrated Farming Systems (IFS)** models by agricultural universities and ICAR institutes has showcased the viability of this combined approach for **income enhancement, resource conservation, and climate resilience**.

Common in smallholder farms in Bihar, Uttar Pradesh, and northeastern states.

- **Main features:** Inclusion of poultry birds (chicken, duck) along with crops and livestock.
- **Livelihood support:** Poultry provides quick returns, meat and eggs for nutrition, and income diversification for rural households.

6. Agroforestry-Based Farming System

The **agroforestry-based farming system** is a sustainable land-use approach that integrates **trees, crops, and/or livestock** on the same piece of land, either in a spatial or temporal sequence. In India, agroforestry has been practiced for centuries in traditional forms such as **home gardens, boundary plantations, and shifting cultivation**. However, with growing recognition of its environmental and economic benefits, agroforestry is now being promoted as a modern and scientifically managed system that contributes to **sustainable agriculture, livelihood security, and climate resilience**.

Agroforestry systems are designed to optimize the productive and protective functions of trees while also supporting agricultural productivity. Trees improve **soil fertility** through nitrogen fixation, organic matter addition, and improved water retention. They also offer **shade, windbreaks, erosion control**, and serve as a habitat for beneficial organisms. From an economic perspective, trees provide **timber, fuelwood, fodder, fruits, nuts, and medicinal products**, thereby diversifying and stabilizing farm income over the long term.

In India, agroforestry is practiced in **varied agro-ecological zones**:

- In **Himalayan regions**, alder-based agroforestry is common in Nagaland and Arunachal Pradesh, where trees are grown alongside maize or millets.
- In the **Indo-Gangetic plains**, farmers plant poplar or eucalyptus trees with wheat, sugarcane, or paddy.
- In **semi-arid regions** of Rajasthan and Gujarat, khejri (Prosopis cineraria) trees are maintained with pearl millet and legumes.
- In **tropical regions** of Kerala and Tamil Nadu, multi-tier home gardens include coconut, banana, spice crops, and timber trees along with vegetables and livestock.

The agroforestry-based farming system is **environmentally sound and economically viable**. It provides **long-term returns from trees** while allowing for short- and medium-term gains from crops and livestock. It enhances **carbon sequestration**, helping mitigate climate change, and contributes to **biodiversity conservation**. In areas prone to drought and land degradation, agroforestry acts as a **buffer system**, protecting crops from extreme weather and restoring soil health. Socially, this system supports **livelihood security** by creating year-round employment and reducing migration. Tree-based products such as bamboo, neem, moringa, and tamarind have strong market potential and are often harvested and processed by **women and small enterprises**, promoting **gender equity** and rural entrepreneurship.

However, agroforestry systems face challenges such as **long gestation periods** for tree crops, **land tenure issues**, lack of market access for tree-based products, and limited technical knowledge among farmers. Farmers may also hesitate to adopt tree planting due to **fear of reduced crop yields** or legal restrictions on felling certain tree species. To overcome these barriers, the Government of India launched the **National Agroforestry Policy (2014)**—the first of its kind globally—which aims to institutionalize agroforestry through research, capacity building, market development, and regulatory reforms. The **Sub-Mission on Agroforestry (SMAF)** under the **National Mission for Sustainable Agriculture (NMSA)** further supports agroforestry adoption through financial and technical assistance. Research institutions like **ICAR-Central Agroforestry Research Institute (CAFRI)** and various **State Agricultural Universities (SAUs)** are working to develop region-specific agroforestry models that suit local climates, soils, and socio-economic conditions. Agroforestry is also being promoted under schemes like **MGNREGA, RKVY**, and **Green India Mission**, linking it to climate action, employment, and ecological restoration.

Prevalent in tribal areas and hilly regions like Chhattisgarh, Jharkhand, and the North-East.

- **Main features:** Combination of trees (timber, fruit, fodder) with crops or pasture.
- **Contribution:** Provides fodder, firewood, fruits, and timber; enhances soil fertility and biodiversity; long-term livelihood security.

7. Integrated Farming System (IFS)

The **Integrated Farming System (IFS)** is a holistic and synergistic approach to agriculture that combines different agricultural enterprises such as **crops, livestock, fisheries, poultry, horticulture, agroforestry, apiculture, sericulture, and mushroom cultivation** within a farm to optimize the use of resources and maximize farm productivity, income, and sustainability. It is especially relevant for **small and marginal farmers in India**, helping them diversify their income sources, enhance food and nutritional security, and increase resilience to climatic and economic shocks. IFS focuses on **recycling**

farm residues and by-products from one component into inputs for another, thus minimizing waste and external inputs. For example, crop residues can be used as fodder or compost, animal manure can enrich soil fertility, and pond water from fish farming can be used for irrigation. This **resource recycling** is central to the ecological and economic viability of IFS.

India, with its diverse agro-climatic zones, supports a wide range of IFS models that vary according to region, climate, and resource availability. Common combinations include:

- **Crop + Livestock + Poultry** (widespread across rainfed areas),
- **Crop + Fish + Duck farming** (popular in Eastern India),
- **Crop + Horticulture + Apiculture** (suitable for hilly and peri-urban regions),
- **Crop + Dairy + Vermicomposting** (ideal for irrigated areas and peri-urban settings),
- **Agroforestry + Crop + Goatery** (suited to arid and semi-arid zones).

One of the major benefits of IFS is **risk minimization**. While mono-cropping can lead to total income loss during crop failure, diversified systems ensure that other enterprises such as poultry or dairy continue to generate revenue. IFS also provides **year-round employment**, promotes **balanced nutrition**, and ensures **better use of available land, labor, water, and capital**. From an environmental perspective, IFS is **eco-friendly and sustainable**. It helps in improving soil health, conserving water, reducing chemical use, and enhancing biodiversity on the farm. The integration of legumes, green manure crops, and organic waste utilization in composting or biogas production contributes to **soil rejuvenation** and **carbon sequestration**.

IFS are particularly significant in the context of **climate change** and **declining farm productivity**. With increasing pressure on land and water, especially in rainfed and tribal areas, IFS offers a sustainable pathway to maintain or improve productivity without degrading natural resources. However, despite its benefits, adoption of IFS faces some constraints. These include **lack of awareness**, **limited access to credit and infrastructure**, and the need for **multi-skill training** for farmers to manage different enterprises. Additionally, marketing support and value chain integration for products like dairy, fish, and horticulture need strengthening.

The Indian Council of Agricultural Research (ICAR), through **All India Coordinated Research Projects on Integrated Farming Systems (AICRP-IFS)**, has developed and validated several region-specific models for different zones of India. Many **Krishi Vigyan Kendras (KVKs)** and **State Agricultural**

Universities (SAUs) have demonstration units and offer training programs to farmers on IFS practices. Government programs such as **Rashtriya Krishi Vikas Yojana (RKVY)**, **National Mission for Sustainable Agriculture (NMSA)**, and **Mahatma Gandhi National Rural Employment Guarantee Scheme (MGNREGS)** are promoting IFS as part of rural development and climate-smart agriculture strategies. The **Integrated Farming System** represents a **scientific, sustainable, and inclusive approach** to agriculture that optimizes productivity, ensures livelihood security, and supports ecological balance. Its widespread adoption can transform Indian agriculture by making it more resilient, resource-efficient, and economically viable, especially for smallholders and rainfed farmers.

Region-Specific Examples of IFS Models

Region	IFS Model	Highlights
Eastern India	Rice + Fish + Duck	Efficient use of waterlogged areas; ducks eat pests and weeds
Southern India	Coconut + Banana + Dairy + Vermicompost	Nutrient recycling, year-round income
Central India	Millets + Goats + Agroforestry (Neem, Subabul)	Drought resilience, fodder from trees
North Eastern Hills	Maize + Piggery + Poultry + Fruits	High-value livestock, low input, local breeds
Western India	Pearl millet + Sheep + Agroforestry	Suited to arid zones; uses Prosopis or Acacia trees
Himalayan region	Barley + Horticulture + Beekeeping	Adds pollination benefits and honey income

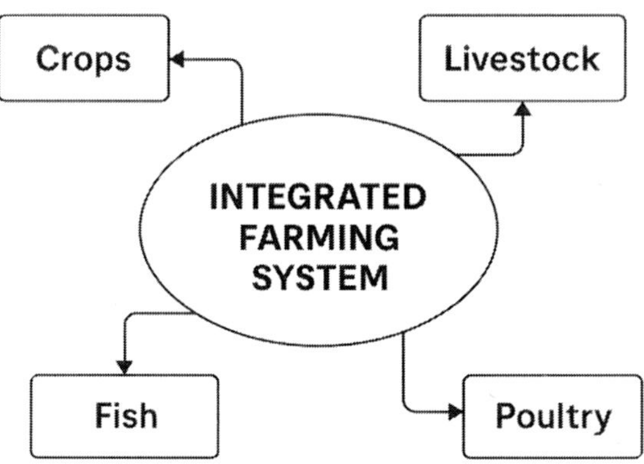

Fig. Cyclical flow of resources in an IFS

Promoted by ICAR and state agricultural universities across India as a sustainable livelihood model.

- **Main features:** Integration of crops, livestock, fishery, poultry, apiculture, biogas, and vermicomposting in a single unit.
- **Livelihood benefits:** Maximizes resource recycling, reduces risk, enhances income stability, and promotes year-round employment.

8. Crop-Sheep/Goat Farming System

The **Crop–Sheep/Goat Farming System** is an important integrated farming approach widely adopted in **arid, semi-arid, and rainfed regions** of India. This system combines **field crop cultivation** with the rearing of **small ruminants**—particularly **sheep and goats**—to maximize land use efficiency, diversify income sources, and enhance the resilience of farming households, especially **small and marginal farmers**. Sheep and goats are well-suited to **low-input systems** and **fragile ecosystems**, as they can thrive on **natural pastures, crop residues, and agro-forest byproducts**. They require less capital, space, and water compared to large animals like cattle or buffalo, making them an ideal component in integrated systems for **resource-poor farmers**.

In this system, **crops are grown primarily for human consumption**, and their residues (like straw, husks, and stubbles) are fed to goats and sheep. In return, animals provide **manure**, which is rich in nutrients and enhances soil fertility when applied back to crop fields. This **mutually beneficial cycle** increases farm sustainability while reducing dependence on external inputs such as chemical fertilizers.

Regional Examples in India:

- In **Rajasthan and Gujarat**, farmers cultivate **millets, legumes, and oilseeds** along with **Barbari or Marwari goats** and local sheep breeds like **Malpura or Marwari sheep**.
- In **Karnataka and Andhra Pradesh**, **sorghum, groundnut, and pigeon pea** are grown along with goats such as **Osmanabadi or Black Bengal** and sheep breeds like **Deccani**.
- In **central tribal belts** (Madhya Pradesh, Chhattisgarh), **coarse grains and pulses** are combined with goat farming, often under **community pasture-based systems**.

The major benefits of this system include:

- **Income diversification**: Crop yields are seasonal, but small ruminants can be sold during lean periods or emergencies.
- **Low risk and quick returns**: Goats and sheep mature faster and require less investment.

- **Manure management**: Improves soil fertility, organic content, and structure.
- **Nutritional security**: Provides milk, meat, and sometimes hide/wool for family use and market sale.

Moreover, **women and landless laborers** often manage small flocks, contributing to their economic empowerment and involvement in farm decision-making. Despite its benefits, this system faces constraints like:

- **Limited access to veterinary care**, vaccinations, and improved breeds;
- **Overgrazing and pasture degradation**;
- **Market fluctuations and inadequate infrastructure** for meat, milk, and skin processing.

To support the adoption and sustainability of the crop–small ruminant system, the government and institutions like **ICAR-Central Sheep and Wool Research Institute**, **National Livestock Mission**, and **KVKs** are promoting:

- Improved breeds (e.g., crossbreeding with Jamunapari, Sirohi);
- Stall-feeding systems using crop byproducts;
- Community pasture management;
- Market linkage for live animals and goat meat ("chevon").

9. Plantation-Based Farming System

The **Plantation-Based Farming System** refers to the cultivation of **commercial tree crops** on a **large scale**, typically for **export or industrial processing**. These systems are characterized by **monoculture of perennial crops**, often in **humid tropical and subtropical regions** of India, and are capital- and labor-intensive in nature. Common crops grown under plantation systems include **tea, coffee, rubber, coconut, arecanut, oil palm, and cashew**.

This system is particularly prominent in the **Western Ghats, Eastern Himalayas, and parts of northeastern India**, where the climate, rainfall, and soil conditions are favorable for such long-duration crops. In many cases, plantation crops form the economic backbone of entire regions and communities.

Key Features

- **Monoculture of perennials**: One major crop is grown over large areas for several years (e.g., tea estates).
- **High capital investment**: Establishment and management involve substantial costs for planting material, irrigation, processing, and marketing.
- **Labor-intensive operations**: Regular tasks include pruning, weeding, harvesting, and processing, often requiring skilled and unskilled labor.
- **Long gestation period**: Returns on investment take time, as most crops take 3–7 years to begin yielding economic produce.
- **Integration with processing**: Crops like tea, coffee, and rubber require immediate processing, necessitating onsite or nearby facilities.

Major Plantation Crops in India and Their Regions

Crop	Major States/Regions	Climatic Requirements
Tea	Assam, West Bengal (Darjeeling), Tamil Nadu	High humidity, moderate temperature, good rainfall
Coffee	Karnataka, Kerala, Tamil Nadu	Shady conditions, moderate altitude, rainfall
Rubber	Kerala, Tripura, Karnataka	Hot and humid, high rainfall
Coconut	Kerala, Tamil Nadu, Karnataka, Andhra Pradesh	Coastal, humid areas
Arecanut	Karnataka, Kerala, Assam	Humid tropics, well-drained soil
Oil Palm	Andhra Pradesh, Telangana, Mizoram	Warm, humid with irrigation support
Cashew	Maharashtra, Goa, Odisha, Andhra Pradesh	Coastal belt, slightly drier climate

Livelihood and Economic Significance

Plantation farming plays a crucial role in providing **employment**, particularly in **rural and tribal areas**. These crops contribute significantly to **India's export earnings**, especially tea, coffee, and spices. Large plantation estates often include **worker housing, schools, and healthcare**, acting as micro-economies.

Environmental and Social Aspects

While plantation crops can stabilize soils and reduce erosion (as in hilly areas), **monoculture practices** may lead to biodiversity loss, soil fertility decline, and heavy use of inputs like fertilizers and pesticides. There are also concerns regarding **labor rights**, especially in older estate systems.

To address these, sustainable practices such as:

- **Agroforestry models** (e.g., shade trees in coffee plantations),
- **Organic certification**,
- **Rainwater harvesting and soil conservation**, and
- **Fair-trade systems**

10. Urban and Peri-Urban Farming Systems

Urban and Peri-Urban Farming Systems (UPFS) refer to the practice of cultivating, processing, and distributing food in and around cities and towns. These farming systems play an increasingly important role in ensuring **food security, nutrition, environmental sustainability**, and **livelihood support** for urban populations. In India, with the rapid pace of urbanization and increasing demand for fresh and safe food, UPFS is gaining attention in both policy and practice.

Urban Farming

Urban farming typically takes place within the limits of a city and involves:

- **Rooftop gardening** (vegetables, herbs, fruit trees),
- **Balcony or terrace farming** using containers or grow bags,
- **Hydroponics and aquaponics** in controlled environments,
- **Vertical farming** using stacked layers or walls.

This form of agriculture is often **small-scale and technology-driven**, primarily for self-consumption or local sale. In cities like **Delhi, Mumbai, Bengaluru, Kolkata, and Hyderabad**, urban dwellers, institutions, and housing societies have embraced urban farming for **fresh produce, cleaner environments, and mental well-being**.

Peri-Urban Farming

Peri-urban areas are zones located on the **fringes of cities**, often transitioning from rural to urban use. Here, farming is typically **semi-commercial or commercial**, and includes:

- **Vegetable and fruit production** for city markets,
- **Dairy farming and poultry** for daily supply chains,

- **Flower cultivation and nurseries**,
- **Mushroom and medicinal plant cultivation**,
- **Fodder production** for urban dairies.

These farms are vital for supplying **perishable and high-value products** to urban centers with **minimum transportation time**, ensuring freshness and reducing carbon footprint.

Key Features and Advantages

- **Proximity to consumers**: Enables rapid marketing and reduced post-harvest losses.
- **Efficient use of space**: Makes use of unused rooftops, backyards, and urban waste spaces.
- **Waste recycling**: Urban farms often use composted kitchen waste or recycled water.
- **Employment generation**: Especially for women, migrants, and urban youth.
- **Climate resilience**: Enhances local food systems during disruptions (e.g., pandemics, supply chain failures).

Challenges

- **Land availability and tenure insecurity** in expanding cities.
- **Water scarcity and pollution** in urban areas.
- **Lack of policy support and extension services** tailored to urban contexts.
- **Limited access to organic inputs, seeds, and marketing support.**

Government and Institutional Support

Several initiatives, both public and private, are promoting urban and peri-urban farming through:

- **Smart City Missions and urban agriculture policies**,
- **ICAR and KVK demonstrations** on rooftop gardens and hydroponics,
- **NGO and startup innovations** in vertical farming, aquaponics, and organic kits.

Cities like **Bhubaneswar, Pune, and Hyderabad** are setting examples with integrated urban farming models, composting programs, and citizen engagement.

Urban and Peri-Urban Farming Systems are emerging as essential components of sustainable urban development in India. They offer a **holistic solution to**

urban food needs, environmental management, and **livelihood support**, while contributing to **nutrition, climate action**, and **urban resilience**. With supportive policies, awareness, and innovation, these systems can be scaled to create **greener, self-reliant cities** in the future. India's diverse farming systems not only ensure **national food security** but also serve as the **backbone of rural livelihoods**. Each system reflects an adaptation to local conditions and resource constraints. The promotion of region-specific, integrated, and sustainable farming systems is vital to improve rural incomes, reduce poverty, and build climate-resilient agriculture.

QUESTION BANK

Short Answer Questions

1. What are farming systems?
2. Name two major farming systems prevalent in India.
3. How does subsistence farming contribute to rural livelihoods?
4. What is mixed farming?
5. Name one advantage of organic farming.
6. How does livestock farming support rural households?
7. What is the main difference between intensive and extensive farming?
8. How does agroforestry benefit farmers?
9. What role does plantation farming play in the Indian economy?
10. Why is diversification important in farming systems?

Multiple Choice Questions (MCQs)

1. Which farming system is primarily practiced for self-consumption?
   - a) Commercial farming
   - b) Subsistence farming
   - c) Plantation farming
   - d) Organic farming
2. Which of the following is an example of commercial farming?
   - a) Growing crops for family use
   - b) Large-scale production of cash crops for sale
   - c) Keeping a home garden
   - d) Practicing slash-and-burn farming

3. Which of these is a key characteristic of mixed farming?
   a) Only crops are grown
   b) Both crops and livestock are raised together
   c) Crops are grown only in winter
   d) Land is left fallow every year
4. What is the main feature of organic farming?
   a) Use of synthetic fertilizers
   b) Avoidance of chemical pesticides and fertilizers
   c) Dependence on genetically modified crops
   d) Heavy use of irrigation
5. In which farming system are tea, coffee, and rubber grown extensively?
   a) Mixed farming
   b) Plantation farming
   c) Subsistence farming
   d) Horticulture
6. Which region in India is known for shifting cultivation?
   a) North-Eastern states
   b) Punjab
   c) Rajasthan
   d) Gujarat
7. Which farming system helps in maintaining soil fertility and biodiversity?
   a) Monocropping
   b) Agroforestry
   c) Commercial farming
   d) Subsistence farming
8. Intensive farming is characterized by:
   a) High use of inputs and resources per unit land
   b) Low productivity and minimal input usage
   c) Leaving land fallow for years
   d) Use of only natural irrigation
9. Rainfed farming is mostly practiced in:
   a) Punjab and Haryana
   b) Rajasthan and Madhya Pradesh
   c) Kerala and Tamil Nadu
   d) Sikkim and Arunachal Pradesh

10. Which farming system integrates trees with crops and livestock?

    a) Commercial farming b) Plantation farming

    c) Agroforestry d) Organic farming

True/False Questions

1. Subsistence farming is primarily focused on producing food for sale. **(False)**
2. Plantation farming is commonly used for cash crops like tea, coffee, and rubber. **(True)**
3. Agroforestry combines agriculture with tree cultivation. **(True)**
4. Intensive farming requires fewer inputs and resources per unit of land. **(False)**
5. Organic farming heavily depends on chemical fertilizers. **(False)**
6. Shifting cultivation is widely practiced in the Indo-Gangetic Plains. **(False)**
7. Livestock farming plays an essential role in the livelihood of small farmers. **(True)**
8. Mixed farming involves growing multiple crops on the same land. **(False)** *(It includes both crops and livestock.)*
9. Commercial farming focuses on large-scale production for market sale. **(True)**
10. Rainfed farming is dependent on artificial irrigation. **(False)** *(It relies on natural rainfall.)*

8

Types of Traditional and Modern Farming Systems

Types of Traditional and Modern Farming Systems

India, with its diverse agro-climatic conditions, has witnessed the coexistence of both **traditional** and **modern farming systems**. These systems reflect the **evolution of agricultural practices**—from indigenous knowledge and subsistence farming to technology-driven, market-oriented approaches.

A. Traditional Farming Systems

Traditional farming systems are based on **indigenous knowledge, ecological balance**, and **subsistence needs**. These systems are often practiced by **tribal, marginal, and small-scale farmers** and are well-adapted to local environments.

Types of Traditional Farming Systems

1. Subsistence Farming

Subsistence farming is a traditional farming system in which a farmer grows crops and/or rears livestock **mainly to meet the needs of their own family**, with little or no surplus for sale in the market. It is generally practiced on **small landholdings**, using **simple tools, indigenous seeds, and family labor**.

- Focus: Meeting the food needs of the family.
- Crops: Millets, pulses, vegetables.
- Tools: Manual or animal-drawn implements.

2. Shifting Cultivation (Jhum)

Shifting cultivation, also known as **Jhum cultivation** in India, is a traditional agricultural practice where farmers **clear a patch of forest land**, **burn the vegetation**, and **cultivate crops for a few years** until the soil fertility declines. After that, the land is **abandoned (left fallow)** and the farmer moves to a new patch of land, allowing the old one to regenerate naturally.

- Practiced in: North-Eastern states and hilly areas.
- Involves: Clearing forest patches, burning biomass, and cultivating for 1–2 years before moving.
- Environmentally sensitive if not managed sustainably.

3. Mixed Cropping

Mixed cropping is a farming practice in which **two or more crops are grown simultaneously on the same piece of land**, without any distinct row arrangement. The main objective is to **minimize the risk of total crop failure** due to pest attacks, diseases, or adverse weather conditions, and to **maximize land use efficiency**.

- Growing two or more crops together on the same field.
- Purpose: Risk reduction and better land use.

4. Rainfed Farming

Rainfed farming is a type of agriculture that **entirely depends on natural rainfall** for meeting the water requirements of crops, without the use of any **artificial irrigation systems**. It is commonly practiced in **regions where irrigation infrastructure is limited or unavailable**, especially in **semi-arid and dry zones**.

- Relies on rainfall without irrigation support.
- Found in: Arid and semi-arid zones.
- Crops: Sorghum, pulses, oilseeds.

5. Pastoral and Nomadic Herding

Pastoral and nomadic herding is a traditional livestock-based livelihood system where communities **depend primarily on the rearing of animals** such as cattle, sheep, goats, camels, or yaks for **milk, meat, wool, and transport**. In **nomadic herding**, the herders **move from one place to another** in search of **fresh grazing pastures and water**, following seasonal patterns.

- Dependent on livestock rearing and seasonal migration.
- Practiced by tribes in Rajasthan, Gujarat, and Himalayan regions.

6. Baranaja System (Uttarakhand)

The **Baranaja system** is a **traditional mixed cropping** practice followed by farmers in the **hilly regions of Uttarakhand**, where **twelve or more crop species** are grown **simultaneously on the same field** during the **Kharif season**. The word *Baranaja* is derived from the Hindi words *"bara"* (twelve) and *"anaj"* (grains), though it may include **more than twelve crops**, including grains, pulses, vegetables, and oilseeds.

- A unique intercropping system involving up to 12 crops.
- Enhances biodiversity, soil health, and food security.

B. Modern Farming Systems

Modern systems are driven by **scientific methods, high input use**, and **commercial orientation**. These aim to increase **productivity, efficiency**, and **market competitiveness**.

Types of Modern Farming Systems

1. Monocropping

Monocropping is the agricultural practice of **growing only one type of crop on the same piece of land, season after season or year after year**. It is a **modern, intensive farming method** widely used for commercial agriculture, especially for crops like **wheat, rice, sugarcane, cotton, and maize**.

- Cultivating a single crop over a large area.
- Used in commercial agriculture (e.g., wheat in Punjab, cotton in Gujarat).
- Risky under changing climate and pest pressures.

2. Mechanized Farming

Mechanized farming is a modern agricultural practice where **machines and equipment** are used extensively for performing various farm operations such as **ploughing, sowing, harvesting, threshing, irrigation, spraying, and transportation**. It aims to **increase efficiency, reduce labor costs**, and **boost agricultural productivity**.

- Use of tractors, harvesters, seed drills, etc.
- Reduces labor dependency and increases operational efficiency.

3. High-Input Farming (Green Revolution)

High-Input Farming, also known as **Green Revolution farming**, refers to the **intensive agricultural practice** that relies heavily on **modern inputs** like **high-yielding varieties (HYVs) of seeds**, **chemical fertilizers**, **pesticides**, and **irrigation** to achieve **maximum crop productivity**, especially for **staple crops** such as **wheat and rice**.

- Emphasis on HYV seeds, chemical fertilizers, and irrigation.
- Led to significant yield gains, especially in rice and wheat.

4. Precision Agriculture

Precision agriculture is a modern, technology-driven farming approach that uses **data, sensors, GPS, drones, and analytics** to **monitor and manage**

crops and soil conditions with high accuracy. The goal is to **optimize input use** (like water, fertilizers, and pesticides), **maximize yields**, and ensure **sustainability and environmental protection**.

- Uses technology (GPS, sensors, drones) for real-time data and targeted inputs.
- Improves resource use efficiency and productivity.

5. Organic Farming

Organic farming is an agricultural system that relies on **natural inputs and ecological processes** to grow crops and raise livestock, avoiding the use of **synthetic chemicals, fertilizers, pesticides, genetically modified organisms (GMOs), and growth hormones**. It emphasizes **soil health, biodiversity, and sustainability**, aiming to produce food in an environmentally and socially responsible way.

- Avoids synthetic inputs; relies on compost, green manure, biofertilizers.
- Promoted for health, environment, and export markets.

6. Integrated Farming System (IFS)

Integrated Farming System (IFS) is a **sustainable agricultural approach** that integrates **different farm components** such as **crops, livestock, poultry, fishery, agroforestry, and biogas units** on a single farm to **maximize resource use efficiency**, **enhance productivity**, and ensure **economic and ecological sustainability**. The outputs (wastes) of one component serve as inputs for another, creating a **closed-loop system**.

- Combines crops, livestock, aquaculture, and forestry.
- Enhances sustainability and income diversification.

7. Hydroponics and Vertical Farming

Hydroponics is a method of **growing plants without soil**, using **nutrient-rich water solutions** to deliver essential minerals directly to the plant roots. It allows for **controlled environment agriculture (CEA)**, where temperature, light, and humidity are optimized for plant growth.

Vertical Farming is a modern farming technique where **crops are grown in vertically stacked layers or towers**, often indoors, using **hydroponics, aeroponics, or aquaponics**. It maximizes **space efficiency**, especially in **urban and peri-urban areas**, and enables **year-round production** with minimal land use.

- Soilless, controlled-environment farming, suitable for urban areas.
- Space-efficient, water-saving, and year-round production.

8. Protected Cultivation (Greenhouses/Polyhouses)

Protected cultivation refers to the practice of growing crops in **controlled environment structures** like **greenhouses, polyhouses, or net houses**, where factors such as **temperature, humidity, light, and ventilation** are regulated to ensure **optimal plant growth** and **high-quality yields**. It is commonly used for **vegetables, flowers, and high-value crops**, protecting them from adverse weather, pests, and diseases.

- Controlled climate for high-value crops like flowers and vegetables.
- Reduces pest risk and enhances quality.

While **traditional farming systems** ensure ecological balance, resilience, and local food security, **modern systems** aim to enhance productivity and profitability through science and technology. A **blended approach**, which incorporates the strengths of both, is crucial for **sustainable agricultural development** in India. India has a diverse range of farming systems due to its varied climate, soil types, and cultural practices. These systems play a crucial role in the livelihood of a significant portion of the population. Here are some of the prevalent farming systems in India contributing to livelihoods:

QUESTION BANK

Short Answer Questions

1. What is the main difference between traditional and modern farming systems?
2. Name two types of traditional farming systems.
3. What is shifting cultivation?
4. How does organic farming differ from conventional farming?
5. What are the main features of modern farming systems?
6. Give one example of a precision farming technique.
7. What is hydroponics?
8. How does mechanization improve modern farming?
9. What role do genetically modified (GM) crops play in modern agriculture?
10. Name one advantage and one disadvantage of traditional farming.

Multiple Choice Questions (MCQs)

1. Which of the following is a traditional farming method?

   a) Hydroponics b) Shifting cultivation

   c) Precision farming d) Vertical farming

2. Which technique is commonly used in modern farming?
   a) Use of synthetic fertilizers and pesticides
   b) Slash-and-burn agriculture
   c) Natural irrigation systems
   d) Crop rotation with only traditional seeds
3. Which farming system is based on soil-less cultivation?
   a) Organic farming
   b) Hydroponics
   c) Mixed cropping
   d) Subsistence farming
4. What is the main objective of precision farming?
   a) Reduce dependence on technology
   b) Maximize efficiency through data-driven decisions
   c) Increase the use of human labor
   d) Depend on shifting cultivation techniques
5. Which of the following is a feature of traditional farming?
   a) Use of high-yield variety (HYV) seeds
   b) Heavy reliance on chemical fertilizers
   c) Dependence on rain and natural fertilizers
   d) Use of genetically modified crops
6. Which modern farming technique uses satellite and drone technology?
   a) Subsistence farming
   b) Precision agriculture
   c) Mixed farming
   d) Agroforestry
7. What is monoculture farming?
   a) Growing multiple crops on the same land
   b) Raising crops and livestock together
   c) Growing only one type of crop over large areas
   d) Farming without any machinery
8. Which farming system involves growing crops without soil?
   a) Agroforestry
   b) Hydroponics
   c) Mixed farming
   d) Terrace farming

9. Which of the following is a **disadvantage** of modern farming?
   a) Increased crop yields
   b) High dependence on chemicals
   c) Reduced mechanization
   d) Lesser use of irrigation
10. What is a **major challenge** of traditional farming?
   a) Use of chemical fertilizers
   b) Low productivity due to outdated techniques
   c) High labor costs due to mechanization
   d) Use of genetically modified crops

True/False Questions

1. Traditional farming relies on natural fertilizers like manure and compost. **(True)**
2. Hydroponics is a traditional farming system. **(False)**
3. Shifting cultivation is a modern farming technique. **(False)**
4. Organic farming avoids synthetic fertilizers and pesticides. **(True)**
5. Precision agriculture makes use of data and technology to improve farming efficiency. **(True)**
6. Traditional farming methods generally have lower yields compared to modern methods. **(True)**
7. Mechanized farming increases labor requirements. **(False)** *(It reduces labor needs.)*
8. Terrace farming is an example of a modern farming technique. **(False)** *(It is a traditional method used in hilly regions.)*
9. Monoculture farming improves biodiversity. **(False)** *(It reduces biodiversity.)*
10. Modern farming techniques can lead to soil degradation if not managed properly. **(True)**

9

Components of Farming System Farming-Based Livelihood Systems- Crops and Cropping Systems

Farming systems or farming-based livelihood systems consist of various components that work together to produce food and generate income for farming households. Understanding crops and cropping systems is a fundamental aspect of this broader framework. A **farming system** or **farming-based livelihood system** is a complex and dynamic arrangement of various interrelated components that work together to support agricultural productivity, sustainability, and livelihood security. The components of such systems are selected and managed based on ecological conditions, socio-economic status, resource availability, and cultural practices. These components not only serve individual purposes but also integrate with one another to enhance overall system efficiency and sustainability.

The **core component** of any farming system is **crop production**, which includes cereals, pulses, oilseeds, vegetables, fruits, and other field crops. The choice of crops is influenced by soil type, rainfall, climate, and market access. Cropping systems such as mono-cropping, mixed cropping, intercropping, crop rotation, and sequential cropping are employed to optimize land use, reduce pest and disease risks, and improve soil fertility.

Livestock is another crucial component that supports the farming system through the provision of milk, meat, manure, and draft power. Cattle, buffaloes, goats, sheep, and poultry are commonly reared. Integration of livestock with crops ensures recycling of nutrients and diversification of income sources.

Fisheries, particularly in water-abundant or coastal regions, form an important supplementary activity. In inland areas, fish farming is integrated with rice cultivation (e.g., rice-fish farming) or practiced in farm ponds, contributing to dietary protein and income.

Horticulture, encompassing the cultivation of fruits, vegetables, spices, and flowers, is becoming increasingly significant in both rural and peri-urban farming systems. It provides high-value produce, enhances nutritional security, and creates employment opportunities.

Agroforestry is another component, where trees are planted along with crops or on field boundaries. It contributes to soil conservation, provides fodder and fuelwood, enhances carbon sequestration, and supports biodiversity.

Poultry farming is widely practiced, even in resource-poor households, as it requires minimal land and investment. It provides regular income and high-quality protein through eggs and meat.

Beekeeping, **mushroom cultivation**, **sericulture**, and other **allied enterprises** are also integrated based on regional suitability. These help diversify income, make use of available biomass and residues, and promote rural entrepreneurship.

Natural resource management, including soil, water, and biodiversity conservation, forms the ecological foundation of a sustainable farming system. Organic farming, conservation agriculture, water harvesting, and use of renewable energy enhance the resilience and long-term productivity of the system. The components of farming-based livelihood systems are designed to complement each other, enhance resource use efficiency, diversify risk, and generate multiple streams of income, ultimately ensuring **economic viability**, **ecological sustainability**, and **social equity** for rural communities.

Here's an overview of the components:

1. Crops

Crops are the primary products cultivated in farming systems. They can be categorized based on various criteria such as:

- **Type of Crop**
    - **Food Crops:** Crops grown primarily for human consumption (e.g., cereals, legumes, fruits, and vegetables).
    - **Cash Crops:** Crops grown for sale and profit (e.g., cotton, coffee, tobacco).
    - **Forage Crops:** Grown primarily for livestock feed (e.g., alfalfa, clover).
- **Growth Duration**
    - **Annuals:** Crops that complete their life cycle in one growing season (e.g., wheat, maize).
    - **Perennials:** Crops that live for multiple years and produce over several seasons (e.g., fruit trees, perennial grasses).
- **Climate and Location**
    - **Tropical Crops:** Adapted to warm temperatures (e.g., rice, cassava).
    - **Temperate Crops:** Thrive in cooler climates (e.g., barley, potatoes).

2. Cropping Systems

Cropping systems refer to the management practices involved in growing crops, including the arrangement and timing of crop plantings. Key aspects include:

- **Monocropping:** Growing a single crop species over a large area for consecutive seasons. This may simplify management and mechanization but can lead to soil degradation and pest issues.
- **Intercropping:** Growing two or more crops together in the same field. This can enhance resource use efficiency, reduce pest pressure, and stabilize yields.
- **Crop Rotation:** Alternating different crops in the same field across seasons to improve soil fertility, reduce disease and pest prevalence, and increase biodiversity.
- **Agroforestry:** Integrating trees and shrubs into farming systems. It can enhance biodiversity, improve soil health, and provide additional income from tree products.
- **Cover Cropping:** Growing crops primarily to cover the soil rather than for harvest, which helps in preventing soil erosion, improving soil organic matter, and managing weeds.

3. Soil Management

Effective soil management practices are critical in cropping systems to maintain soil fertility and structure, which could include:

- **Organic Practices:** Using compost, manure, and green manures to enhance soil health.
- **Conventional Practices:** Use of chemical fertilizers and pesticides to boost crop production, although careful management is critical for environmental sustainability.

4. Water Management

Water availability and management practices significantly impact cropping systems:

- **Irrigation Systems:** Techniques like drip, sprinkler, or flood irrigation to ensure adequate water supply.
- **Rainfed Systems:** Relying on natural rainfall, which may involve selecting drought-resistant crop varieties.

5. Economic Considerations

The economic viability of crops and cropping systems is vital for the sustainability of farming systems, including:

- **Market Access:** Proximity to potential buyers, availability of transportation, and the presence of local markets can influence crop selection.
- **Cost of Inputs:** Labor, fertilizers, seeds, and equipment costs play a crucial role in determining what and how much to grow.

6. Sociocultural Factors

The choice of crops and cropping systems may also be influenced by:

- **Cultural Preferences:** Traditional diets and local customs affecting crop selection.
- **Labor Availability:** Skills and labor supply in the local context.

The components of cropping systems are interconnected, and their successful management requires understanding the biological, physical, economic, and social aspects of farming systems. Sustainable practices within these systems can help enhance productivity while also conserving environmental resources. When designing farming systems, it is crucial to consider the ecological balance, economic viability, and social acceptance of the crops and practices chosen.

QUESTION BANK

Short Answer Questions

1. What is meant by a cropping system?
2. Name two types of cropping systems commonly practiced in India.
3. What is crop rotation?
4. Define intercropping with an example.
5. What is the main advantage of mixed cropping?
6. How does monocropping impact soil fertility?
7. Name one sustainable cropping system that improves soil health.
8. What is relay cropping?
9. How does crop diversification benefit farmers?
10. What is the difference between multiple cropping and mixed cropping?

Multiple Choice Questions (MCQs)

1. Which of the following is an example of **monocropping**?
   a) Growing rice and wheat together
   b) Cultivating only maize in a large field for years
   c) Growing multiple crops in the same season
   d) Planting legumes and cereals alternately
2. In which cropping system are different crops grown on the same land in **sequential seasons**?
   a) Mixed cropping b) Intercropping
   c) Crop rotation d) Monocropping
3. **Intercropping** involves:
   a) Growing the same crop over large areas
   b) Growing two or more crops together in rows
   c) Alternating crops every two years
   d) Leaving the land fallow for one season
4. Which cropping system is beneficial for pest control and soil fertility improvement?
   a) Monocropping b) Crop rotation
   c) Shifting cultivation d) Continuous cropping
5. **Mixed cropping** is practiced mainly to:
   a) Increase yield stability and reduce crop failure risks
   b) Grow only cash crops
   c) Improve export quality of crops
   d) Increase soil degradation
6. Which of the following is an **example of multiple cropping**?
   a) Growing wheat and rice in alternate years
   b) Growing wheat, maize, and pulses in the same field in one season
   c) Growing only sugarcane on a farm
   d) Growing crops in rotation every decade
7. **Alley cropping** involves:
   a) Growing crops between rows of trees
   b) Leaving the land fallow for one season
   c) Growing a single crop in an entire field
   d) Rotating crops every 10 years

8. What is the major drawback of monocropping?
   a) Increased soil fertility
   b) Higher pest and disease risk
   c) Better crop yield stability
   d) Sustainable nutrient cycling

9. Which cropping system is best suited for rainfed agriculture?
   a) Monocropping
   b) Intercropping
   c) High-input commercial farming
   d) Hydroponics

10. Relay cropping refers to:
   a) Growing two or more crops in a sequence with overlapping growth periods
   b) Growing crops in different fields each season
   c) Leaving fields without crops for one year
   d) Growing only one crop per year

True/False Questions

1. Monocropping reduces soil fertility over time. **(True)**
2. Intercropping helps in reducing pest attacks naturally. **(True)**
3. Crop rotation involves growing the same crop on the same land every year. **(False)** *(Different crops are grown in sequential seasons.)*
4. Mixed cropping is mainly done to maximize market profit rather than reduce risks. **(False)** *(It is done to reduce risks of crop failure.)*
5. Alley cropping integrates trees with agricultural crops. **(True)**
6. Multiple cropping means growing different crops in different seasons on separate lands. **(False)** *(It means growing different crops in the same field within a year.)*
7. Monocropping promotes biodiversity in the field. **(False)** *(It reduces biodiversity.)*
8. Crop diversification can improve farmers' income and soil health. **(True)**
9. Relay cropping and intercropping are the same. **(False)** *(Relay cropping has overlapping growing periods.)*
10. Traditional farming methods often rely on mixed and intercropping to maintain soil fertility. **(True)**

10

Livestock (Dairy, Piggery, Goatry Poultry, Duckry)

Livestock integration into a farming system offers multiple benefits, including manure for soil fertility, animal products for food and income, and utilization of crop residues. Livestock components like dairy, piggery, goatry, poultry, and duckry are essential for **enhancing income, improving livelihoods, and ensuring sustainability** within farming systems. They are particularly important in **diversified and integrated farming approaches**, enabling **multi-output farming** that benefits rural communities both economically and nutritionally. **Livestock** plays a vital role in farming systems across India and serves as a **major source of livelihood**, especially for small and marginal farmers. It provides **nutrition, income, employment**, and **farm inputs** such as manure and draft power, while also serving as a **risk buffer** in times of crop failure.

- **Dairy:** Dairy farming involves the rearing of cows and buffaloes for **milk production**. It is a prominent source of **regular income** in rural areas and supports food security through **milk and milk products**. Cow dung is also used as **organic manure** and fuel.
  - **Definition:** Raising cattle (cows, buffaloes) primarily for milk production.
  - **Components**
    - Animal Breed: Selecting breeds with high milk yield and adaptability.
    - Feeding: Providing balanced nutrition through fodder, concentrates, and supplements.
    - Housing: Providing proper shelter and sanitation.
    - Health Management: Disease prevention and treatment.
    - Milking and Processing: Efficient and hygienic milking practices and processing of milk into products like cheese, yogurt, and butter.

  - Manure Management: Using manure as fertilizer or biogas production.
- **Integration:** Dairy can be integrated with crop production by using crop residues as feed and manure to fertilize fields.

- **Piggery:** Piggery is common in **northeastern states and tribal areas**. Pigs are reared for **meat production** and are highly prolific and fast-growing. They require **low input** and adapt well to diverse climates.
  - **Definition:** Raising pigs for meat production.
  - **Components**
    - Breed Selection: Choosing breeds with good growth rates and meat quality.
    - Feeding: Providing a balanced diet, often utilizing kitchen waste and crop by-products.
    - Housing: Providing appropriate shelter and sanitation to prevent disease.
    - Health Management: Vaccination and disease control.
    - Waste Management: Managing pig manure to minimize environmental impact.
  - **Integration:** Pig manure can be used as fertilizer or in biogas production. Pigs can also be fed with crop residues and by-products.
- **Goatry:** Goat rearing, often called the "**poor man's cow**," is suitable for **resource-poor farmers**. Goats provide **meat, milk, and manure**, and they thrive in **arid and semi-arid regions** where other livestock may not survive.
  - **Definition:** Raising goats for meat, milk, and fibre (e.g., cashmere).
  - **Components**
    - Breed Selection: Choosing breeds suited to the local environment and production goals.
    - Feeding: Providing grazing land and supplemental feed.
    - Housing: Providing shelter from the elements and predators.
    - Health Management: Disease prevention and parasite control.
    - Manure Management: Utilizing goat manure for soil improvement.
  - **Integration:** Goats can graze on land unsuitable for crops, and their manure can be used to improve soil fertility.

- **Poultry:** Poultry farming includes the rearing of **chickens for eggs and meat**. It is a fast-growing sector in India and a **good source of daily income**, particularly for **landless and smallholder farmers**.
  - **Definition:** Raising birds (chickens, turkeys, ducks, and geese) for meat and/or eggs.
  - **Components**
    - Breed Selection: Choosing breeds based on desired product (meat or eggs).
    - Feeding: Providing balanced feed formulations.
    - Housing: Providing appropriate housing to protect from predators and weather.
    - Health Management: Vaccination and disease control.
    - Egg Collection and Handling: Proper collection, cleaning, and storage of eggs.
    - Waste Management: Managing poultry litter and manure.
  - **Integration:** Poultry manure is a valuable fertilizer. Poultry can also be integrated into orchards or other cropping systems to control pests and weeds.
- **Duckry:** Duck farming is suitable for **wetland and rice-growing areas**, especially in eastern and southern India. Ducks require **less feed**, can forage naturally, and are resistant to many diseases.
  - **Definition:** Raising ducks for meat and eggs.
  - **Components**
    - Breed Selection: Choosing breeds for meat or egg production.
    - Feeding: Ducks can forage for food and also require supplemental feed.
    - Housing: Ducks need access to water for swimming and cleaning.
    - Health Management: Disease prevention.
    - Egg Collection and Handling: Proper collection, cleaning, and storage of eggs.
    - Waste Management: Managing duck manure.
  - **Integration:** Ducks can be integrated into rice paddies to control pests and weeds and fertilize the soil with their droppings.

III. Factors Influencing Farming System Choice:

- **Agro-climatic conditions:** Temperature, rainfall, soil type, and altitude.

- **Market demand:** Consumer preferences and prices for different products.
- **Availability of resources:** Land, water, labour, capital, and technology.
- **Farmer's knowledge and skills:** Experience and expertise in different farming practices.
- **Government policies and support:** Subsidies, extension services, and research programs.
- **Socio-economic factors:** Family size, access to credit, and cultural practices.

IV. Sustainability Considerations

- A sustainable farming system aims to:
- Maintain or improve soil health.
- Conserve water resources.
- Minimize pollution.
- Enhance biodiversity.
- Reduce reliance on external inputs.
- Increase resilience to climate change.
- Improve the livelihoods of farmers.

By carefully considering these components and their interactions, farmers can develop farming systems that are both productive and sustainable.

QUESTION BANK

Short Answer Questions

1. What are the major livestock farming systems in India?
2. Name two high-yielding dairy cattle breeds in India.
3. What is the primary purpose of pig farming (piggery)?
4. Which livestock farming system is known as the "poor man's cow"?
5. How does poultry farming contribute to rural livelihoods?
6. What is the main advantage of duck farming (duckry) over poultry farming?
7. Name one common disease in poultry farming.
8. Which breed of goat is best for milk production in India?
9. What are the major products obtained from pig farming?
10. Why is livestock farming important for small and marginal farmers?

Multiple Choice Questions (MCQs)

1. Which breed is famous for high milk production in India?

a) Gir b) Yorkshire

c) Kadaknath d) Large White Pig

2. Which of the following livestock is commonly referred to as the "poor man's cow"?

a) Cattle b) Goat

c) Pig d) Duck

3. Which livestock farming system produces eggs and meat?

a) Dairy farming b) Poultry farming

c) Piggery d) Goatry

4. Which of the following is a fast-growing breed of pig in India?

a) Sahiwal b) Large White Yorkshire

c) Black Bengal Goat d) Aseel

5. Which livestock system is commonly practiced in wetland areas?

a) Poultry farming b) Duck farming

c) Goat farming d) Pig farming

6. Which of these livestock species has the highest feed conversion efficiency?

a) Goat b) Pig

c) Cow d) Sheep

7. Which Indian state is the leading producer of poultry meat and eggs?

a) Punjab b) Tamil Nadu

c) Assam d) Gujarat

8. Which goat breed is mainly raised for meat production?

a) Jamunapari b) Beetal

c) Black Bengal d) Sahiwal

9. Which of the following livestock produces "duck eggs," which are richer in fat than chicken eggs?

a) Poultry b) Duck

c) Goat d) Pig

10. Which of the following diseases affects dairy cattle?

a) Foot and Mouth Disease
b) Ranikhet Disease
c) Swine Fever
d) Newcastle Disease

True/False Questions

1. Cattle farming is mainly done for milk and draft power. **(True)**
2. Pig farming is less profitable than dairy farming. **(False)** *(Piggery has high feed efficiency and profitability.)*
3. Poultry farming provides both eggs and meat. **(True)**
4. Duck farming is not suitable for wetland areas. **(False)** *(Duck farming is best suited for wetlands.)*
5. Goat farming requires large amounts of land for grazing. **(False)** *(Goats can be raised in small areas and require less feed.)*
6. Swine fever is a common disease in poultry farming. **(False)** *(It affects pigs, not poultry.)*
7. Ducks can feed on aquatic plants and insects, reducing feed costs. **(True)**
8. Broiler chickens are raised mainly for egg production. **(False)** *(Broilers are raised for meat.)*
9. The Gir and Sahiwal breeds are known for high milk yield. **(True)**
10. Livestock farming contributes significantly to rural employment. **(True)**

11

Horticultural Crops, Agro - Forestry Systems, Aquaculture

These three components—**horticultural crops, agroforestry systems, and aquaculture**—play a significant role in **diversifying agriculture**, **increasing income**, and **promoting sustainability** in farming-based livelihood systems across India. Horticultural crops, agroforestry systems, aquaculture, duck/ poultry cum fish, and dairy cum - all describe different farming systems, often with a focus on integration and diversification. Here's a breakdown of each:

- **Horticultural Crops:** This refers to the cultivation of fruits, vegetables, flowers, and ornamental plants. It's a broad category encompassing many different specific crops and growing techniques. Horticulture includes the **cultivation of fruits, vegetables, flowers, spices, condiments, and medicinal plants**. These crops are **high-value**, labor-intensive, and contribute significantly to **nutritional security**, **income generation**, and **employment creation**, especially for smallholders and women. **Examples:** Mango, banana, guava, brinjal, tomato, onion, rose, marigold, turmeric.

Advantages

- High market demand and export potential.
- Year-round employment opportunities.
- Improves diet diversity and nutrition.

Relevance: Especially important in **peri-urban areas**, **home gardens**, and **tribal regions**.

- **Agroforestry Systems:** This is an integrated land-use system that combines trees and shrubs with agricultural crops and/or livestock. Agroforestry aims to create more sustainable and productive land-use practices by utilizing the beneficial interactions between trees and other components. Examples include alley cropping (planting crops between rows of trees), silvopasture (integrating trees and grazing livestock), and forest farming (growing crops under a forest canopy). **Agroforestry** is a land-use system that integrates **trees and shrubs with crops and/or**

livestock on the same unit of land. It offers **ecological and economic benefits**, enhances **resilience to climate change**, and improves **soil fertility and water conservation**.

Types of Agroforestry

- **Agri-silviculture:** Crops + trees
- **Silvi-pasture:** Trees + pasture/animals
- **Agri-horti-silviculture:** Crops + horticulture + trees

Benefits

- Provides timber, fuelwood, fodder, fruits, and shade.
- Protects soil against erosion.
- Enhances biodiversity and carbon sequestration.

Examples: Poplar + wheat system in Punjab; Mango + legumes + boundary trees in Odisha.

Table: Comparison of Horticultural Crops, Agroforestry Systems, and Aquaculture

Component	Description	Examples	Key Benefits	Suitable Regions
Horticultural Crops	Cultivation of fruits, vegetables, spices, flowers, and medicinal plants	Mango, Tomato, Onion, Rose, Turmeric	High income, nutritional security, year- round employment	Peri-urban areas, tribal zones, kitchen gardens
Agroforestry Systems	Integration of trees with crops and/or livestock	Wheat + Poplar, Mango + Legumes + Teak trees	Soil conservation, timber/fodder/fuel wood, climate resilience	Semi-arid, hilly, degraded, and rainfed areas
Aquaculture	Farming of fish and other aquatic organisms in water bodies	Rohu, Catla, Tilapia, Prawns	High protein yield, efficient water use, income diversification	Eastern India, coastal belts, water-abundant zones

- **Aquaculture:** This involves the farming of aquatic organisms such as fish, crustaceans, mollusks, and aquatic plants. It can be done in fresh water, brackish water, or saltwater environments. Aquaculture is the **cultivation of aquatic organisms** such as **fish, prawns, and crabs** under controlled or semi-controlled conditions. It is a highly productive and profitable activity, particularly in areas with access to water bodies.

Types

- **Inland aquaculture** (ponds, tanks)
- **Brackish water aquaculture** (coastal areas)
- **Integrated systems** (e.g., rice–fish farming)

Benefits

- High protein food source.
- Efficient use of water resources.
- Enhances income and nutrition.

Popular Species: Rohu, Catla, Tilapia, Prawns.

Regions: West Bengal, Andhra Pradesh, Odisha, Assam.

- **Duck/Poultry cum Fish Farming:** This is an integrated farming system where ducks or poultry are raised in conjunction with fish farming. The ducks/poultry provide manure that fertilizes the fishpond, promoting algae growth and providing food for the fish. The fish, in turn, can help control pests in the duck/poultry area. This system reduces the need for external inputs and improves overall efficiency.
- **Dairy cum Farming:** "Cum" in this context means "together with" or "in addition to." Dairy cum farming means integrating dairy farming with other agricultural activities. This could involve combining dairy production with crop cultivation (using manure as fertilizer, growing feed crops), poultry farming, or other livestock. The goal is to create a more diversified and sustainable farm operation.

These systems often aim to improve resource utilization, reduce waste, increase productivity, and enhance overall farm sustainability. They reflect a move towards more integrated and ecological approaches to agriculture.

COMPONENTS OF FARMING-BASED LIVELIHOOD SYSTEMS

DESCRIPTION	EXAMPLES	KEY BENEFITS	SUITABLE REGIONS
HORTICULTURAL CROPS	Cultivation of fruits, vegetables, spices, flowers, and medicinal plants	High income, nutritional security, year-round employment	Peri-urban areas, tribal zones, kitchen gardens
AGROFORESTRY SYSTEMS	Integration of trees with crops and/or livestock	Soil conservation, timber/fodder/ fuelwood, climate resilience	Semi-arid, hilly, degraded, and rainfed areas
AQUACULTURE	Farming of fish and other aquatic organisms in water bodies	High protein yield, efficient water use income diversification	Eastern India, coastal belts. water-abundant zones

Fig. Comparison of Horticultural Crops, Agroforestry Systems, and Aquaculture

QUESTION BANK

Short Answer Questions

1. What are horticultural crops?
2. Name two major fruit crops grown in India.
3. Define agroforestry.
4. What are the benefits of integrating trees into farmland?
5. Name two common agroforestry systems practiced in India.
6. What is aquaculture?
7. How does aquaculture contribute to food security?
8. Name two commonly farmed fish species in India.
9. What is the difference between inland and marine aquaculture?
10. How does agroforestry help in soil conservation?

Multiple Choice Questions (MCQs)

1. Which of the following is an example of a horticultural crop?

a) Rice b) Mango

c) Wheat d) Maize

2. Which horticultural crop is known as the "King of Fruits"?

a) Banana b) Mango

c) Apple d) Pineapple

3. Which of the following is NOT an agroforestry system?

a) Silvopastoral system b) Agrisilviculture

c) Monoculture farming d) Agrihorticulture

4. What is the main objective of agroforestry?

a) Only increase crop production

b) Integrate trees with crops and livestock for sustainability

c) Replace all agricultural crops with trees

d) Convert farmland into forests

5. Which of the following is a major advantage of agroforestry?

a) Decreases soil fertility

b) Reduces biodiversity

c) Provides additional income sources

d) Increases soil erosion

6. Which of the following is a common fish species used in aquaculture?

a) Rohu b) Wheat

c) Cotton d) Rubber

7. What is mariculture?

a) Farming of marine organisms in open oceans

b) Cultivating crops in saline soil

c) Raising poultry in coastal regions

d) Growing horticultural crops underwater

8. Which method of aquaculture involves raising fish in artificial tanks?

a) Cage culture b) Hydroponics

c) Aeroponics d) Desert farming

9. Which is a major constraint in aquaculture?

a) Unlimited water resources

b) High risk of disease outbreaks

c) No need for feeding

d) No government regulations

10. Which of the following trees is commonly used in agroforestry?

a) Eucalyptus b) Sugarcane

c) Rice d) Sunflower

True/False Questions

1. Horticultural crops include fruits, vegetables, and ornamental plants. **(True)**
2. Agroforestry does not help in biodiversity conservation. **(False)** *(It promotes biodiversity.)*
3. Mango and guava are common fruit crops grown in India. **(True)**
4. Agroforestry systems can improve soil fertility. **(True)**
5. Silvopastoral systems combine trees with crop production. **(False)** *(They integrate trees with livestock grazing.)*
6. Aquaculture is the farming of fish, crustaceans, and aquatic plants. **(True)**
7. Mariculture refers to fish farming in freshwater ponds. **(False)** *(It involves marine aquaculture.)*

8. Agroforestry practices can help in climate change mitigation. **(True)**
9. In polyculture aquaculture, only one species of fish is raised. **(False)** *(Multiple species are raised together.)*
10. Integrated farming systems can combine aquaculture with livestock and crop farming. **(True)**

12

Duck/Poultry Cum Fish, Dairy Cum Fish, Piggery Cum Fish etc.

It seems that you're interested in integrated farming systems involving aquaculture combined with other types of livestock. Integrated farming systems (IFS) are agricultural systems that combine different agricultural practices to maximize productivity, enhance sustainability, and diversify income sources. Here's a brief overview of how these combinations work:

1. **Duck/Poultry cum Fish:** This system integrates **duck or poultry rearing with fish culture** in a common water body. Ducks or chickens roam over or near fish ponds, and their **droppings directly fertilize the pond**, increasing natural fish food (plankton).

- **Overview**: This system involves raising ducks or poultry alongside fish in a shared environment, often in ponds.
- **Benefits**
    - **Nutrient Cycling**: Duck or poultry manure can supply nutrients to the water, enhancing fish growth.
    - **Pest Control**: Ducks help control pests in rice paddies while providing additional protein sources.
    - **Water Quality**: The movement of ducks or poultry can improve the aeration of water and help manage algae.
- **Considerations**: It's essential to manage feeding and water levels to prevent over-fertilization and maintain good water quality.

2. **Dairy cum Fish:** In this system, **cattle dung** from dairy animals is used to **fertilize the fish pond**, promoting plankton growth which serves as fish feed. It also allows for **nutrient recycling** and integration of livestock with aquaculture.

- **Overview**: In this system, dairy cattle are raised along with fish in a pond or tank.
- **Benefits**

- **Manure Utilization**: Dairy manure can be used to fertilize fish ponds, promoting algae growth that serves as feed for fish.
- **Reduced Waste**: Instead of just disposing of manure, it contributes to aquatic productivity.
- **Diversification**: Farmers can sell both milk and fish, providing a more stable income.

- **Considerations**: Management of feed types and ensuring proper ratios in feeding are critical to maintain a healthy environment for both fish and cows.

3. **Piggery cum Fish:** Piggery–fish integration is common in **northeast India**. Pig sheds are constructed beside or above the pond, and pig waste serves as a **rich organic input** for fish ponds.

- **Overview**: This integrated system involves raising pigs and fish, typically utilizing the pig manure as fertilizer in fish ponds.
- **Benefits**
  - **Waste-to-Resource**: Pig waste is rich in nutrients that benefit fish growth.
  - **Economic Synergy**: Selling both pork and fish creates multiple revenue streams.
  - **Food Security**: This system can improve food security for the farm and surrounding community.
- **Considerations**: Care must be taken to manage waste runoff and water quality to avoid pollution and ensure a balanced ecosystem.

General Advantages of Integrated Farming Systems

- **Resource Efficiency**: Maximizes the use of available resources and minimizes waste.
- **Resilience**: Diversifies income and reduces reliance on a single crop or livestock type.
- **Sustainability**: Enhances ecological balance and can reduce the environmental impact of farming.

General Considerations

- **Water Management**: Essential to ensure balanced water quality for both fish and livestock.
- **Disease Management**: Integrated systems can facilitate the spread of diseases across species groups, so management practices should be robust.

- **Community Involvement**: Such systems can benefit from community knowledge and resource sharing.

Integrated farming systems require careful planning and management to ensure that all components work synergistically. Implementing a correct setup can lead to improved agricultural productivity and sustainability.

QUESTION BANK

Short Answer Questions

1. What is an integrated farming system?
2. How does poultry farming benefit fish culture in an integrated system?
3. What are the advantages of combining pig farming with fish farming?
4. Name two benefits of a dairy cum fish farming system.
5. How do ducks contribute to fish farming in a duck cum fish system?
6. What types of fish are commonly raised in integrated farming systems?
7. Why is pig manure useful in fish farming?
8. How does integrating dairy farming with fish farming improve resource efficiency?
9. What are the environmental benefits of integrated livestock-fish farming?
10. Name one major challenge faced in integrated livestock-fish farming systems.

Multiple Choice Questions (MCQs)

1. **What is the main purpose of integrated livestock-fish farming?**
   a) To increase waste production
   b) To improve resource utilization and profitability
   c) To eliminate fish farming
   d) To focus only on livestock farming

2. **Which livestock species is commonly integrated with fish farming due to its high manure output?**
   a) Sheep b) Pigs
   c) Horses d) Camels

3. **In a poultry cum fish farming system, what do the fish primarily benefit from?**
   a) Poultry feathers
   b) Droppings as organic manure
   c) Noise from the poultry farm
   d) Egg production

4. **Which fish species is best suited for integrated livestock-fish farming?**
   a) Rohu b) Trout
   c) Salmon d) Tuna

5. **What is the role of ducks in a duck cum fish farming system?**
   a) They provide extra protein for fish
   b) They act as natural aerators for ponds
   c) They eat all the fish
   d) They produce milk for fish

6. **Which of the following is NOT an advantage of integrated farming systems?**
   a) Higher efficiency in resource use
   b) Increased income diversification
   c) High chemical fertilizer usage
   d) Better waste recycling

7. **Which livestock waste is the richest in nutrients for fish farming?**
   - a) Poultry droppings
   - b) Cow dung
   - c) Pig manure
   - d) Goat droppings
8. **Which of the following systems is best for small-scale farmers to maximize land productivity?**
   - a) Dairy cum fish
   - b) Poultry cum fish
   - c) Piggery cum fish
   - d) All of the above
9. **Which integrated farming system is most commonly practiced in Southeast Asia?**
   - a) Poultry cum fish
   - b) Dairy cum fish
   - c) Pig cum fish
   - d) Duck cum fish
10. **Why is dairy farming combined with fish farming?**
    - a) To use cow dung as manure for fish ponds
    - b) To allow cows to drink pond water
    - c) To breed cows in water
    - d) To make fish feed from milk

True/False Questions

1. Integrated farming systems help in waste recycling. **(True)**
2. Poultry droppings can serve as organic fertilizer for fish ponds. **(True)**
3. Pigs are not suitable for integrated fish farming. **(False)** *(Pig manure is highly beneficial for fish growth.)*
4. Duck farming does not contribute to fish farming. **(False)** *(Ducks provide natural aeration and pest control.)*
5. Dairy cows directly feed fish in an integrated system. **(False)** *(Only their manure is used in ponds.)*
6. Fish raised in integrated systems have lower productivity than those in traditional systems. **(False)** *(They benefit from organic manure and have higher productivity.)*
7. Integrated livestock-fish farming reduces production costs for farmers. **(True)**
8. The combination of pigs and fish farming is considered environmentally unsustainable. **(False)** *(If managed properly, it can be highly sustainable.)*

9. Duck manure helps in the growth of plankton, which is a natural fish feed. **(True)**
10. Integrated farming systems promote better land and water use efficiency. **(True)**

13

Small Medium and Large Enterprises Including Value Chains

Small, medium, and large enterprises (SMEs and LEs) play a vital role in any economy by providing goods, services, and employment opportunities. Each category has unique characteristics, challenges, and significance within the economy. Value chains are integral to understanding how these enterprises operate and compete in the market. Enterprises in agriculture and allied sectors are classified based on **investment, turnover, and scale of operation**. These enterprises form the **backbone of agri-based value chains**, providing crucial support in **production, processing, storage, marketing**, and **distribution**. Here's an overview of SMEs, LEs, and their value chains:

Small Enterprises

Definition

- Typically have fewer than 50 employees.
- Annual revenue varies by country but is generally less than $10 million.

Characteristics

- Often family-owned or local businesses.
- Serve niche markets and local communities.
- Limited resources for marketing, technology, and personnel.
- Higher flexibility and capacity for innovation due to fewer bureaucratic hurdles.

Challenges

- Limited access to capital and financing.
- Difficulty in scaling operations.
- Competition from larger firms.
- Regulatory burdens can be disproportionately high.

Medium Enterprises

Definition

- Usually have between 50 to 250 employees.
- Annual revenue varies by country but typically lies between $10 million to $50 million.

Characteristics

- More structured than small enterprises but still maintain some level of agility.
- Broader customer base, often operating at regional or national levels.
- Better access to funding compared to small enterprises but still face challenges.

Challenges

- Balancing operational efficiency with growth aspirations.
- Navigating complex regulatory landscapes.
- Competing with larger corporations that have more resources.

Large Enterprises

Definition

- Generally have over 250 employees.
- Annual revenue exceeds $50 million, with some even reaching billions.

Characteristics

- Highly structured organizations with well-defined hierarchies.
- Significant resources for marketing, research, and development.
- Operate on a global scale, often having a diverse product portfolio.

Challenges

- Maintaining innovation in a bureaucratic structure.
- Responding rapidly to market changes.
- Managing a diverse workforce and potential cultural differences across locations.

Value Chains

Definition

A value chain is a set of activities that a company performs to deliver a valuable product or service to the market. This concept helps businesses analyze their internal processes, understand where they add value, and identify areas for improvement.

Components of a Value Chain

1. **Inbound Logistics:** Receiving, warehousing, and inventory management of raw materials.
2. **Operations:** Processes that transform inputs into final products or services.
3. **Outbound Logistics:** Activities required to get the finished product to the customer, including storage and order fulfillment.
4. **Marketing and Sales:** Strategies and activities to promote and sell products or services.
5. **Service:** Activities related to maintaining the value of the product or service post-sale, such as customer service and repair.

Value Chains in SMEs and LEs

- **Innovation and Flexibility in SMEs:** Small businesses can often pivot and innovate within their value chains more rapidly than larger enterprises due to less bureaucratic red tape. They can quickly adapt to consumer preferences and market conditions.
- **Efficiency in LEs:** Large enterprises leverage economies of scale to optimize their value chains, achieving cost efficiencies in production, distribution, and marketing. They may also use sophisticated supply chain management techniques to enhance collaboration and integration across their value chains.
- **Collaboration across Enterprises:** SMEs can participate in value chains by partnering with larger firms, providing specialized services or products that complement the offerings of larger enterprises. This collaboration can enhance the overall efficiency and competitiveness of the value chain.

Conclusion

Understanding the distinctions between small, medium, and large enterprises, along with the intricacies of their value chains, is crucial for policymakers, business owners, and stakeholders. This knowledge is essential for fostering economic growth, creating jobs, and driving innovation within various sectors of the economy. Businesses can enhance their competitive advantage by effectively managing their value chains and leveraging the unique strengths of their size category.

Agri Value Chain Integration

Value Chain Stage	Enterprise Role
Input Supply	Seed companies, agri-equipment SMEs
Production	Farmer Producer Organizations (FPOs), contract farming
Post-Harvest Handling	Sorting, grading, cleaning (medium units)
Processing	SMEs (basic) and large industries (advanced)
Packaging & Branding	Medium and large-scale enterprises
Marketing & Retail	E-commerce platforms, agri startups, aggregators

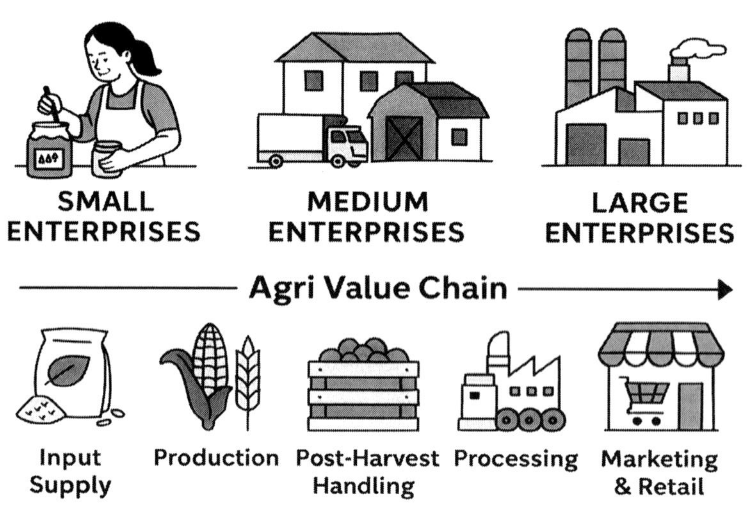

QUESTION BANK

Short Answer Questions

1. What is the definition of a Small and Medium Enterprise (SME)?
2. How are enterprises classified as small, medium, or large?
3. What is a value chain in business?
4. Name two challenges faced by small enterprises.
5. How do large enterprises contribute to economic growth?
6. What are the benefits of value chain integration for businesses?
7. Give an example of a business operating within a value chain.
8. How do SMEs benefit from participating in global value chains?

9. What role does technology play in value chain management?
10. What are the key differences between a small enterprise and a large enterprise?

Multiple Choice Questions (MCQs)

1. Which of the following is a key characteristic of small enterprises?

a) High capital investment

b) Large-scale production

c) Limited employees and lower revenue

d) International market presence

2. Which sector has the highest number of small and medium enterprises?

a) Automotive industry

b) Agriculture and food processing

c) Aerospace industry

d) Oil and gas sector

3. What is the main advantage of participating in a value chain?

a) Higher costs

b) Increased efficiency and profitability

c) Reduced market access

d) Dependence on a single supplier

4. Which factor is commonly used to classify enterprises as small, medium, or large?

a) Number of employees and annual revenue

b) Number of customers

c) Length of time in business

d) Social media presence

5. Which of the following is a key function of a value chain?

a) Increasing costs for businesses

b) Managing supply chain logistics efficiently

c) Reducing business growth

d) Preventing market expansion

6. **Which of the following is an example of a small enterprise?**
   a) A multinational oil company
   b) A local bakery with 5 employees
   c) A global automobile manufacturer
   d) A large telecommunications firm

7. **How do large enterprises benefit from value chains?**
   a) They have higher costs and lower profits
   b) They increase operational efficiency and reduce costs
   c) They become dependent on small businesses
   d) They focus only on one stage of production

8. **What is a supply chain?**
   a) A group of businesses selling the same product
   b) A network involved in producing and delivering a product
   c) A single business with multiple locations
   d) A list of suppliers and customers

9. **Which of the following is NOT a part of a business value chain?**
   a) Manufacturing b) Logistics
   c) Customer service d) Watching TV

10. **Which organization supports the development of SMEs globally?**
    a) World Bank b) NASA
    c) FIFA d) WWE

True/False Questions

1. Small enterprises typically have fewer than 50 employees. **(True)**
2. Large enterprises do not participate in value chains. **(False)** *(They play a major role in value chains.)*
3. The value chain concept applies only to manufacturing businesses. **(False)** *(It applies to all industries.)*
4. Medium-sized enterprises contribute significantly to employment generation. **(True)**
5. SMEs generally have lower access to financial resources than large enterprises. **(True)**

6. The primary goal of a value chain is to increase production costs. **(False)** *(The goal is to improve efficiency and reduce costs.)*
7. Small enterprises cannot be part of a global value chain. **(False)** *(They can participate through exports and partnerships.)*
8. Large enterprises have more complex value chains than small businesses. **(True)**
9. Businesses that integrate value chains can gain a competitive advantage. **(True)**
10. SMEs are more flexible and adaptive to market changes than large enterprises. **(True)**

14

Secondary Enterprises as Livelihood Components for Farmers

Secondary enterprises can play a crucial role in improving the livelihoods of farmers by diversifying their income sources, reducing risks associated with dependence on a single crop or farming activity, and increasing overall economic resilience. In addition to primary farming activities, many farmers engage in **secondary enterprises** to **diversify their income**, **reduce risk**, and **ensure year-round employment**. These activities add value to agricultural produce or provide services linked to the agri-sector, often through **micro-enterprises, cottage industries**, or **village-level entrepreneurship**.

Definition

Secondary enterprises refer to **non-farm or off-farm activities** that are directly or indirectly linked to agriculture, often based on local resources, skills, and traditional knowledge.

Here are several ways that secondary enterprises can serve as effective livelihood components for farmers:

1. Value-Added Products

Farmers can engage in the production of value-added products. This may include processing raw agricultural products into items such as jams, cheeses, dried fruits, or specialty sauces. By doing so, they can enhance the market value of their products and tap into niche markets.

2. Agri-Tourism

Farmers can diversify their income by participating in agri-tourism. This could involve offering farm tours, educational workshops, farm stays, or events like harvest festivals. Agri-tourism not only generates additional income but also helps connect consumers with food sources.

3. Organic Fertilizers and Pesticides

Farmers can produce and sell organic fertilizers and natural pest control solutions, utilizing by-products or waste materials from their primary farming activities. This aligns with sustainable farming practices and meets the increasing consumer demand for organic products.

4. Livestock Farming

Introducing livestock farming can provide an alternative source of income. Farmers can raise poultry, goats, pigs, or cattle, leading to diversified income streams. Selling meat, eggs, or dairy products can enhance a farm's economic viability.

5. Renewable Energy Production

Farmers can invest in renewable energy technologies, such as solar, wind, or biogas systems. Not only can they reduce energy costs, but they can also sell surplus energy back to the grid, creating an additional revenue stream.

6. Crop Rotation and Intercropping

Implementing crop rotation and intercropping systems allows farmers to cultivate a variety of crops on the same land. This practice can improve soil health, reduce pests and diseases, and provide multiple sources of income throughout the year.

7. Direct Sales and Community Supported Agriculture (CSA)

Farmers can sell their products directly to consumers through farmers' markets, online platforms, or subscription-based delivery services like CSA. This reduces dependence on middlemen, increases profit margins, and strengthens community relationships.

8. Handicrafts and Artisan Goods

If farmers possess skills in handicrafts, they can produce and sell handmade goods such as textiles, pottery, or woodworking items. This adds a creative outlet and provides additional income opportunities.

9. Workshops and Training

Farmers can utilize their knowledge to offer workshops or training sessions for other local farmers or community members. Topics may range from sustainable farming practices to cooking classes, providing both educational value and income.

10. Collaboratives and Cooperatives

Joining or forming cooperatives allows farmers to pool resources, share knowledge, and collectively market their products. This can lead to better bargaining power in the marketplace and reduced costs for inputs.

Challenges and Considerations

While establishing secondary enterprises can be beneficial, farmers may face challenges such as resource limitations, lack of access to markets, need for additional skills, and initial financial investment. It's essential to conduct feasibility studies and ensure there is a market demand for the secondary products or services being developed.

Conclusion

Secondary enterprises offer farmers numerous opportunities to enhance their livelihoods, create economic resilience, and contribute to sustainable agricultural practices. Through innovation and improved market access, farmers can leverage their existing resources to create additional income streams and improve their overall well-being.

Table: Examples of Secondary Enterprises for Farmers

Type of Enterprise	Activities
Food Processing Units	Making pickles, jams, juices, papad, ghee, jaggery
Dairy & Animal Products	Ghee, paneer, curd making, wool processing
Poultry & Meat Processing	Dry meat processing, egg trays, feather-based handicrafts
Handicrafts & Rural Art	Bamboo crafts, coir making, basket weaving, mat weaving
Agro-Tourism	Farm stays, demonstration farms, eco-tourism units
Agri-based Services	Seed treatment, spraying services, agri-tool rentals, soil testing
Compost/Bio-input Units	Vermicompost production, bio-fertilizers, organic inputs
Textile & Weaving	Khadi, handloom, cotton spinning in cotton-growing areas
Beekeeping & Honey Units	Honey processing and wax product manufacturing
Mushroom Cultivation	Mushroom farming, processing and value-added mushroom products

Benefits of Secondary Enterprises

- **Income diversification** during off-seasons.
- Reduces **dependency on crop yield**.
- Promotes **rural self-reliance** and entrepreneurship.
- Enhances **value addition** and **marketable surplus**.
- Empowers **women and youth** through small-scale business models.

Support Systems

- **Government Schemes:** FMFME (PM Formalization of Micro Food Processing Enterprises), NRLM, MSME schemes.
- **Institutional Support:** KVKs, NABARD, FPOs, SHGs, cooperatives.

- **Capacity Building:** Skill training, entrepreneurship development programs.

QUESTION BANK

Short Answer Questions

1. What are secondary enterprises in the context of farm-based livelihoods?
2. Name two examples of secondary enterprises that farmers can engage in.
3. How do secondary enterprises contribute to a farmer's income?
4. What are some common challenges faced by farmers in secondary enterprises?
5. How does food processing serve as a secondary enterprise for farmers?
6. What role does agroforestry play in secondary enterprises?
7. How can dairy farming be considered both a primary and a secondary enterprise?
8. Why is rural handicraft production considered a secondary enterprise for farmers?
9. How can farmers benefit from integrating beekeeping with crop farming?
10. What are the government schemes that support farmers in developing secondary enterprises?

Multiple Choice Questions (MCQs)

1. Which of the following is an example of a secondary enterprise for farmers?

a) Growing wheat
b) Dairy processing
c) Land plowing
d) Irrigation setup

2. Which secondary enterprise involves producing goods like jams, pickles, and dairy products?

a) Poultry farming
b) Food processing
c) Crop rotation
d) Irrigation management

3. What is the primary benefit of secondary enterprises for farmers?

a) Increased dependency on middlemen
b) Additional income and employment opportunities
c) Reduced farm productivity
d) Higher input costs

4. **Which of the following secondary enterprises is related to animal husbandry?**
   a) Organic vegetable farming b) Agro-tourism
   c) Dairy processing d) Timber production
5. **Which of the following is NOT a secondary enterprise?**
   a) Beekeeping b) Agroforestry
   c) Growing rice d) Vermicomposting
6. **Which industry is closely linked to secondary enterprises in farming?**
   a) IT industry b) Food processing industry
   c) Construction industry d) Mining industry
7. **Which product is commonly obtained from agroforestry as a secondary enterprise?**
   a) Solar panels b) Medicinal plants
   c) Smartphones d) Cement
8. **Which government initiative in India supports secondary enterprises for farmers?**
   a) PM-KISAN b) National Livelihood Mission
   c) MGNREGA d) Digital India
9. **Why is fish farming considered a secondary enterprise for farmers?**
   a) It requires advanced machinery
   b) It provides an additional source of income
   c) It does not require water
   d) It reduces soil fertility
10. **Which of the following factors is crucial for the success of secondary enterprises?**
    a) Lack of storage facilities b) Strong market linkages
    c) High taxation d) Poor infrastructure

True/False Questions

1. Secondary enterprises help farmers reduce their dependence on seasonal crops. **(True)**
2. Dairy processing is not considered a secondary enterprise. **(False)** *(It is a common secondary enterprise.)*

3. Beekeeping can improve crop yields while providing additional income. **(True)**
4. Agroforestry does not contribute to secondary enterprises. **(False)** *(It provides timber, fuel, and medicinal plants.)*
5. Lack of market access is a major challenge for farmers in secondary enterprises. **(True)**
6. Secondary enterprises have no role in rural employment generation. **(False)** *(They provide significant employment opportunities.)*
7. Food processing is a major secondary enterprise that adds value to agricultural products. **(True)**
8. Government schemes like SFURTI support rural artisans and secondary enterprises. **(True)**
9. Poultry farming cannot be integrated with secondary enterprises. **(False)** *(It can be integrated with egg processing and meat production.)*
10. Secondary enterprises improve farmers' resilience to climate change and economic shocks. **(True)**

15

Factors Affecting Integration of Various Enterprises of Farming for Livelihood

The integration of various farming enterprises—such as crops, livestock, poultry, fishery, horticulture, agroforestry, and others—plays a vital role in strengthening rural livelihoods. However, successful integration depends on a variety of interrelated factors. The first and foremost is the agro-climatic condition of the area, which includes soil type, rainfall pattern, temperature, humidity, and elevation. These factors determine the suitability of different enterprises and their possible combinations. For example, fish farming is more viable in water-abundant regions, while goat farming is more appropriate in dry or hilly regions. Landholding size and its layout also greatly influence enterprise integration. Farmers with larger and well-demarcated plots have more flexibility to integrate different components such as ponds for fish, sheds for livestock, and space for crops or horticulture. On the other hand, marginal and small farmers may find it challenging to accommodate multiple enterprises without careful planning. Similarly, water availability is a key determinant, especially for integrating enterprises like fishery, dairy, and horticulture, which require regular and ample water supply. The presence of water harvesting structures or access to irrigation facilities enhances integration potential.

Financial capital and investment capacity also play a crucial role. Setting up poultry units, building animal shelters, or creating fish ponds requires upfront investment. Therefore, the availability of credit facilities, subsidies, or group-based investments through cooperatives or self-help groups can encourage farmers to diversify. Equally important is the skill level and technical knowledge of farmers. Proper training, exposure visits, and support from Krishi Vigyan Kendras (KVKs), agricultural universities, and extension workers are essential to ensure efficient and sustainable management of integrated systems. Market access and local demand influence the choice and success of enterprises. Farmers tend to integrate those components that are in demand in the nearby markets—for example, milk, eggs, vegetables, or fish. Without a reliable market, even a well-integrated farm system may not yield good income. Labour availability is another practical consideration. Integrated farming systems are labour-intensive and often require continuous supervision. The availability of family labour or the affordability of hired labour determines the scale and complexity of integration.

Infrastructure and input availability, such as veterinary care, feed supply, transportation, and storage, are supportive factors that enhance the sustainability of enterprise integration. Further, policies and institutional support from the government—through schemes like RKVY, PMMSY, or NABARD initiatives—can boost farmer confidence and promote integrated farming systems. Lastly, social and cultural aspects, including traditional practices, community preferences, and religious beliefs, also influence the choice of enterprises. For instance, piggery may be widely accepted in the northeastern states but less so in other regions due to cultural constraints. The integration of farming enterprises for livelihood enhancement depends on a complex interplay of biophysical, economic, social, and institutional factors. A well-planned, location-specific approach that considers these factors can lead to better resource utilization, increased farm income, and sustainable rural development.

The integration of various enterprises within farming for improved livelihoods is influenced by a multitude of factors. These can be categorized into social, economic, environmental, and institutional dimensions. Below are some key factors that can affect the integration of farming enterprises:

Economic Factors

1. **Market Access**: Availability of markets and the ability to access them can significantly influence the success of integrated farming. Proximity to markets can reduce transportation costs and improve profitability.
2. **Capital and Investment**: Availability of financial resources and access to credit can support the establishment and expansion of integrated farming systems.
3. **Input Costs**: The costs of seeds, fertilizers, and other inputs can affect the feasibility and attractiveness of integrating multiple enterprises.
4. **Economies of Scale**: Larger operations may benefit from reduced per-unit costs, encouraging integration to leverage these economies.

Social Factors

1. **Community Dynamics**: Social networks and community support systems can facilitate the sharing of knowledge, resources, and labor, thereby promoting integrated farming.

2. **Cultural Practices**: Traditional practices and beliefs may either support or hinder the adoption of integrated farming methods.
3. **Knowledge and Skills**: Farmers' education levels, training, and experience play crucial roles in their ability to implement and manage integrated systems effectively.

Environmental Factors

1. **Soil Health**: The quality and type of soil can influence which enterprises can be integrated. Healthy soil can support diverse crops and livestock.
2. **Climate**: Weather patterns and climate can affect the viability of various enterprises and their interdependence within an integrated system.
3. **Biodiversity**: A diverse ecosystem can enhance resilience against pests and diseases and improve productivity.

Technological Factors

1. **Access to Technology**: Availability of modern farming equipment and sustainable agricultural technologies can enhance productivity and efficiency in integrated systems.
2. **Innovation**: The adoption of innovative practices and techniques can foster successful integration through increased productivity and conservation of resources.

Policy and Institutional Factors

1. **Government Policies**: Supportive policies, subsidies, and extension services can promote integrated farming practices.
2. **Regulatory Frameworks**: Compliance with regulations relating to health, safety, and environmental sustainability can impact the viability of integrated systems.
3. **Research and Development**: Investment in research to develop appropriate technologies and practices that support integration can help farmers adopt more sustainable methods.
4. Government schemes (like **RKVY, PMMSY, eNAM**, etc.) can facilitate enterprise integration.
5. **FPOs**, **SHGs**, and **cooperatives** play a vital role in resource pooling and service delivery.

Management Practices

1. **Farm Management Skills**: Strong management practices, including planning, monitoring, and decision-making, are essential for successfully integrating various enterprises.

2. **Risk Management**: Understanding and managing risks associated with different farming enterprises can influence integration strategies.

Infrastructure

1. **Transportation**: Good transportation networks facilitate the movement of goods to markets and access to essential inputs.
2. **Irrigation and Water Management**: Availability of water resources and irrigation systems can affect the integration of livestock and crop production.

Access to Information

1. **Extension Services**: Access to agricultural extension services and information can empower farmers with knowledge about integrated practices.
2. **Technology Platforms**: Modern communication technologies (e.g., mobile apps, online resources) can facilitate knowledge sharing and support farmer networks.

By addressing these factors holistically, farmers can improve the integration of various enterprises, leading to more sustainable livelihoods and enhanced productivity.

Fig. Factors affecting integration of various enterprises of farming for livelihood

QUESTION BANK

Short Answer Questions

1. What is enterprise integration in farming?
2. Name two major factors affecting the integration of farming enterprises.
3. How does soil quality impact enterprise integration in farming?
4. Why is water availability crucial for integrating different farming enterprises?
5. How do government policies influence the integration of agricultural enterprises?
6. What role does market access play in the success of integrated farming systems?
7. How can financial resources impact a farmer's ability to integrate multiple enterprises?
8. Why is knowledge and skill level an important factor in enterprise integration?
9. How does climate change affect integrated farming systems?
10. What are the economic benefits of integrating multiple farming enterprises?

Multiple Choice Questions (MCQs)

1. Which of the following is a key factor affecting the integration of farming enterprises?

a) Soil fertility b) Internet speed

c) Fashion trends d) Television advertisements

2. Why is water availability a critical factor in integrated farming systems?

a) It is needed for irrigation and livestock

b) It increases fertilizer costs

c) It reduces labor requirements

d) It makes soil infertile

3. How do government policies affect integrated farming?

a) By providing financial support and subsidies

b) By banning multiple enterprises

c) By increasing tax rates on integrated farming

d) By discouraging innovation

4. **Which of the following is an economic factor that affects enterprise integration?**

   a) Climate conditions b) Market access

   c) Soil structure d) Rainfall patterns

5. **Why is labor availability important for integrating multiple farming enterprises?**

   a) More labor is needed to manage different enterprises

   b) Labor decreases the productivity of enterprises

   c) It makes farming unprofitable

   d) Labor is not needed in integrated farming

6. **Which type of financial support can encourage farmers to integrate enterprises?**

   a) High-interest loans b) Agricultural subsidies

   c) Increased tax rates d) Import duties

7. **Which of the following is a major risk of integrating multiple enterprises?**

   a) Increased market opportunities b) Higher financial investment

   c) Improved land utilization d) Diversified income sources

8. **How does technology impact integrated farming systems?**

   a) It increases efficiency and productivity

   b) It makes farming less profitable

   c) It reduces the need for integration

   d) It eliminates the role of natural resources

9. **Which of the following external factors affects integrated farming enterprises the most?**

   a) Cultural festivals b) Political instability

   c) Personal preferences of farmers d) Social media trends

10. **Why is knowledge and skill level important for enterprise integration?**

    a) Skilled farmers can manage multiple enterprises effectively

    b) More knowledge leads to reduced productivity

    c) Skill level has no impact on farming

    d) Integrated farming does not require expertise

True/False Questions

1. Soil fertility has no impact on the integration of farming enterprises. **(False)** *(Soil fertility determines crop suitability and productivity.)*
2. Farmers with better access to markets are more likely to integrate enterprises. **(True)**
3. Water availability is not a major factor in integrated farming. **(False)** *(Water is essential for irrigation, livestock, and fish farming.)*
4. Government subsidies and financial support help promote integrated farming. **(True)**
5. Climate change has no effect on farming enterprise integration. **(False)** *(Extreme weather conditions can disrupt farming enterprises.)*
6. High labor costs can discourage farmers from integrating multiple enterprises. **(True)**
7. Access to technology and mechanization can improve the efficiency of integrated farming. **(True)**
8. Lack of financial resources does not affect a farmer's ability to integrate enterprises. **(False)** *(Capital is needed to establish and manage multiple enterprises.)*
9. Diversified farming enterprises help reduce income risks for farmers. **(True)**
10. Poor infrastructure, such as roads and storage facilities, has no effect on enterprise integration. **(False)** *(Poor infrastructure limits market access and profitability.)*

16

Feasibility of Different Farming Systems for Different Agro-Climatic Zones

India's vast geographical expanse and diverse climatic conditions have led to the development of distinct agro-climatic zones. These zones, as defined by the Planning Commission and ICAR, provide a scientific basis for selecting location-specific farming systems that ensure sustainable productivity, livelihood security, and efficient resource use. The feasibility of farming systems in these zones depends on factors such as rainfall, temperature, soil type, water availability, cropping intensity, and socio-economic conditions. The feasibility of different farming systems significantly depends on the specific agro-climatic zone, which encompasses factors such as temperature, rainfall, soil type, and elevation.

India's vast and varied geography is divided into 15 agro-climatic zones by the Planning Commission. Each zone has specific soil types, climatic conditions, topography, and resource availability. These factors influence the most feasible and sustainable farming systems. Below is a detailed description of suitable farming systems for each zone.

1. Western Himalayan Region (Jammu & Kashmir, Himachal Pradesh, Uttarakhand)

- **Climate:** Cold temperate to alpine.
- **Feasible Farming Systems:**
  - Horticulture-based (apple, pear, walnut)
  - Agri-horti-pastoral (wheat/barley + fruits + sheep/goat)
  - Apiculture and sericulture
- **Justification:** Hilly terrain and climate favor fruits and small livestock.

2. Eastern Himalayan Region (Arunachal Pradesh, Sikkim, parts of Assam)

- **Climate:** Humid subtropical to alpine.
- **Feasible Farming Systems:**

  - Agroforestry (bamboo, cardamom, orange)
  - Horti-pastoral
  - Traditional shifting cultivation (Jhum)
- **Justification:** Forest cover and rainfall suit mixed systems.

3. Lower Gangetic Plains (West Bengal)

- **Climate:** Hot and humid.
- **Feasible Farming Systems:**
  - Rice-fish-livestock
  - Rice-wheat-vegetable
  - Integrated crop-livestock
- **Justification:** Water-rich, fertile soils support intensive systems.

4. Middle Gangetic Plains (Eastern UP, Bihar)

- **Climate:** Subtropical humid.
- **Feasible Farming Systems:**
  - Rice-wheat-pulses
  - Crop-livestock-dairy
  - Agro-horticulture (mango, guava)
- **Justification:** High cropping intensity and livestock support.

5. Upper Gangetic Plains (Western UP)

- **Climate:** Subtropical.
- **Feasible Farming Systems:**
  - Crop-dairy (wheat, sugarcane + dairy)
  - Agro-horticulture
  - Poultry and vegetable integration
- **Justification:** Fertile land and market access enable diversification.

6. Trans-Gangetic Plains (Punjab, Haryana, Delhi)

- **Climate:** Semi-arid to sub-humid.
- **Feasible Farming Systems:**
  - Crop-livestock (rice-wheat + dairy)
  - Agroforestry (poplar/eucalyptus)
  - Protected cultivation (polyhouses)
- **Justification:** Intensive farming with need for sustainability.

7. Eastern Plateau and Hills (Jharkhand, Chhattisgarh, Odisha parts)

- **Climate:** Hot, sub-humid to humid.
- **Feasible Farming Systems:**
  - Agroforestry with millets and pulses
  - Horti-agriculture-livestock
  - Tribal integrated systems
- **Justification:** Poor soils and tribal economy benefit from integration.

8. Central Plateau and Hills (MP, parts of Rajasthan and Maharashtra)

- **Climate:** Semi-arid to sub-humid.
- **Feasible Farming Systems**
  - Dryland farming (sorghum, pulses)
  - Agro-pastoral
  - Horticulture-livestock
- **Justification:** Rainfed regions need resource-efficient systems.

9. Western Plateau and Hills (Maharashtra, MP parts)

- **Climate:** Semi-arid.
- **Feasible Farming Systems**
  - Cotton-pulses-livestock
  - Agroforestry (fruit trees)
- **Justification:** Low rainfall and hardy crops dominate.

10. Southern Plateau and Hills (Karnataka, Andhra Pradesh, Tamil Nadu parts)

- **Climate:** Tropical.
- **Feasible Farming Systems**
  - Millets-pulses-livestock
  - Agro-horti-livestock
  - Sericulture
- **Justification:** Drought-prone areas require low-input systems.

11. East Coast Plains and Hills (Odisha, Andhra Pradesh, Tamil Nadu coast)

- **Climate:** Humid coastal.
- **Feasible Farming Systems**

- Rice-fish-poultry
- Horti-livestock
- Brackishwater aquaculture

- **Justification:** Coastal and deltaic areas ideal for integrated farming.

12. West Coast Plains and Ghat (Kerala, coastal Karnataka, Goa)

- **Climate:** Hot, humid.
- **Feasible Farming Systems**
  - Horti-aqua (coconut, banana + fish)
  - Spice-based (pepper, nutmeg)
  - Multi-storeyed cropping
- **Justification:** High-value plantations and fisheries dominate.

13. Gujarat Plains and Hills

- **Climate:** Arid to semi-arid.
- **Feasible Farming Systems**
  - Cotton-groundnut-livestock
  - Agroforestry with fodder trees
  - Dryland horticulture
- **Justification:** Resource conservation and hardy crops preferred.

14. Western Dry Region (Rajasthan Desert)

- **Climate:** Arid.
- **Feasible Farming Systems:**
  - Agro-pastoral (pearl millet + goats/camels)
  - Silvipasture
  - Rainwater harvesting-based systems
- **Justification:** Focus on livestock and water conservation.

15. Islands Region (Andaman & Nicobar, Lakshadweep)

- **Climate:** Tropical, humid.
- **Feasible Farming Systems**
  - Coconut + aquaculture + poultry
  - Mixed cropping (tubers, spices)
  - Coastal integrated farming
- **Justification:** Self-sufficiency due to isolation; aquatic resources.

Besides these factors influence the most feasible and sustainable farming systems 15 agro climatic zone of India, we explore the suitability of various farming systems in different agro-climatic zones of the World:

1. Tropical Rainforest Zone

- **Characteristics**: High rainfall, high humidity, and consistent temperatures.
- **Farming Systems**
  - **Shifting Cultivation**: Suitable for subsistence farming; relies on the natural regeneration of forests.
  - **Agroforestry**: Combines agricultural crops with tree planting; promotes biodiversity and sustainability.
  - **Perennial Crops**: Coffee, cacao, and rubber thrive in these conditions.

2. Tropical Savanna Zone

- **Characteristics**: Distinct wet and dry seasons, moderate rainfall.
- **Farming Systems**:
  - **Crop-Livestock Integration**: Mixed farming is common here, combining crops like millet and sorghum with livestock.
  - **Pulses Cultivation**: Incorporating legumes can improve soil health and provide food.
  - **Dairy Farming**: Feasible due to available grasses during rainy seasons.

3. Subtropical Zone

- **Characteristics**: Mild winters and hot summers, sufficient rainfall.
- **Farming Systems**
  - **Horticulture**: Vegetables, fruits, and ornamental plants can be cultivated year-round.
  - **Cotton and Sugarcane**: Favorable for cash crops contributing to local economies.
  - **Aquaculture**: Suitable in areas with sufficient water resources.

4. Temperate Zone

- **Characteristics**: Four distinct seasons with moderate rainfall.
- **Farming Systems**
  - **Dairy and Beef Farming**: Extensive grazing systems can support livestock.

  - **Cash Cropping**: Crops like wheat, corn, and soybeans are well-suited.
  - **Organic Farming**: Increased consumer demand for organic produce encourages this practice.

5. Mediterranean Zone

- **Characteristics**: Hot, dry summers and mild, wet winters.
- **Farming Systems**
  - **Vineyards and Olive Orchards**: Cultivation of grapes for wine and olives for oil.
  - **Citrus Fruits**: Oranges and lemons thrive under these conditions.
  - **Sustainable Grazing**: Utilizing drought-resistant livestock breeds for pasture-based systems.

6. Arid and Semi-Arid Zones

- **Characteristics**: Low rainfall, extreme temperature variations.
- **Farming Systems**
  - **Dryland Farming**: Techniques such as no-till farming and crop rotation can enhance moisture retention.
  - **Solar Agriculture**: Utilizing solar energy to power irrigation systems.
  - **Agroecology**: Focus on sustainable practices that improve soil and water conservation.

7. Highland or Mountain Zones

- **Characteristics**: Cooler temperatures, varying rainfall depending on elevation.
- **Farming Systems**
  - **Terracing**: To cultivate on slopes and prevent soil erosion.
  - **Potato and Barley Cultivation**: Crops that are well-adapted to high altitudes.
  - **Livestock Farming**: Raising sheep and goats, which are well-suited for grazing on rugged terrain.

Important Considerations

1. **Soil Health**: The choice of farming system should focus on maintaining and improving soil health, adaptation to local conditions, and resilience to climate changes.

2. **Water Management**: In arid and semi-arid regions, efficient water use is critical, leading to practices such as drip irrigation and rainwater harvesting.
3. **Biodiversity**: Integrating diverse crops and animal breeds can improve resilience against pests and diseases.
4. **Market Access**: Consideration of market demands and access is crucial for the viability of cash crops in any agro-climatic zone.
5. **Cultural Practices**: Local traditions and practices should be respected and integrated into farming systems to enhance community acceptance and sustainability.

Ultimately, selecting the right farming system for an agro-climatic zone involves a thorough understanding of the local ecological, social, and economic conditions. Collaboration with local farmers, agronomists, and extension services can aid in the effective implementation of sustainable farming practices.

QUESTION BANK

Short Answer Questions

1. Define an agro-climatic zone. Why is this classification important for farming systems in India?
2. What type of farming system is most suitable for the Western Himalayan Region and why?
3. Mention two integrated farming systems feasible in the Lower Gangetic Plains and explain their components.
4. Describe the role of agroforestry in the Eastern Himalayan Region.
5. How do farming systems in the Western Dry Region of Rajasthan address water scarcity?
6. Explain why rice-fish-livestock integration is ideal for the East Coast Plains.
7. Which crops and animals are generally included in the agro-pastoral system of the Central Plateau?
8. Highlight the importance of sericulture in the Southern Plateau and Hills region.
9. How is the farming system in island regions different from the mainland?
10. Suggest two farming systems that can help reduce dependency on groundwater in Punjab and Haryan
11. What is an agro-climatic zone?

12. Name two factors that determine the feasibility of a farming system in an agro-climatic zone.
13. Why is rice cultivation most suitable in humid and high-rainfall areas?
14. How does soil type influence the choice of farming systems in different agro-climatic zones?
15. Which farming system is most suitable for arid and semi-arid regions?
16. What role does temperature play in determining the feasibility of a farming system?
17. Why is integrated farming considered beneficial in tropical agro-climatic zones?
18. How do water availability and irrigation facilities impact the feasibility of farming systems?
19. Which agro-climatic zone is best suited for horticulture-based farming systems?
20. What are some key constraints in implementing farming systems in hilly regions?

Multiple Choice Questions (MCQs)

Choice Questions (MCQ)

1. Which zone is most suited for horticulture-based farming involving apple and pear?

   a) Lower Gangetic Plains b) Western Himalayan Region

   c) Western Dry Region d) Trans-Gangetic Plains

2. The most appropriate farming system for West Bengal's agro-climatic zone is:

   a) Cotton-livestock system b) Rice-fish-livestock

   c) Agro-pastoral d) Millets-based farming

3. Which of the following zones is characterized by dryland horticulture and livestock?

   a) Western Plateau b) Lower Gangetic Plains

   c) Islands Region d) Eastern Himalayan Region

4. The agro-climatic zone most affected by shifting cultivation is:\n

   a) Central Plateau \n b) Trans-Gangetic Plains

   c) Eastern Himalayan Region \n d) Upper Gangetic Plains \n

5. Which integrated farming system is most suitable for coastal Odisha?
   a) Agro-pastoral b) Rice-fish-poultry
   c) Cotton-groundnut d) Wheat-sugarcane
6. The Western Dry Region of Rajasthan primarily practices:
   a) Agro-horticulture
   b) Silvipasture and camel-based systems
   c) Rice-fish farming
   d) Sericulture
7. The zone where multi-storeyed cropping and spice cultivation is common is:\n
   a) West Coast Plains and Ghat b) Gujarat Plains
   c) Central Plateau d) Lower Gangetic Plains
8. Which of these zones has potential for brackish water aquaculture?
   a) Eastern Plateau and Hills b) Islands Region
   c) Upper Gangetic Plains D) Central Plateau
9. A common component in the farming systems of the Southern Plateau is:
   a) Fish farming b) Cotton cultivation
   c) Sericulture d) Jhum cultivation
1. Agro-climatic zoning helps in:
   a) Dividing land by soil color
   b) Increasing crop prices
   c) Selecting suitable farming systems
   d) Importing agricultural inputs **Answer:** C
2. Which of the following is a primary factor affecting the feasibility of a farming system in an agro-climatic zone?
   a) Local cuisine preferences b) Soil type and climate
   c) Road infrastructure d) Political stability
3. Which farming system is most suitable for arid and semi-arid regions?
   a) Rainfed mixed farming b) Paddy-based farming
   c) Agroforestry d) Wetland farming

4. What makes the Indo-Gangetic plains ideal for wheat and rice cultivation?
   a) High elevation
   b) Sandy soil
   c) Fertile alluvial soil and good irrigation facilities
   d) Low temperature throughout the year
5. Which farming system is best suited for coastal agro-climatic zones?
   a) Dryland agriculture
   b) Aquaculture and fish farming
   c) Cattle ranching
   d) Agroforestry
6. Why is organic farming more feasible in temperate agro-climatic zones?
   a) High pest resistance
   b) Stable temperature and moderate rainfall
   c) High fertilizer use
   d) Lower soil fertility
7. Which type of farming is commonly practiced in mountainous agro-climatic zones?
   a) Plantation farming
   b) Shifting cultivation and terrace farming
   c) Large-scale mechanized farming
   d) Rice-wheat cropping
8. What is the most limiting factor for farming in arid and desert agro-climatic zones?
   a) Excess rainfall
   b) Poor soil fertility
   c) Water scarcity
   d) Lack of labor
9. Which region in India is most suitable for tea and coffee plantation farming?
   a) Indo-Gangetic Plains
   b) Western Ghats and Northeast India
   c) Thar Desert
   d) Deccan Plateau

10. Which farming system is best suited for flood-prone areas?

    a) Dairy farming b) Poultry farming

    c) Floating agriculture d) Agroforestry

11. Why is livestock farming important in drought-prone regions?

    a) It requires large amounts of water

    b) It is less dependent on rainfall than crop farming

    c) It decreases soil fertility

    d) It is unsuitable for dry regions

True/False Questions

1. Different agro-climatic zones require different farming systems for optimal productivity. **(True)**
2. Paddy-based farming is the most suitable system for desert agro-climatic zones. **(False)** *(Paddy requires high water availability, which deserts lack.)*
3. Agroforestry is a feasible farming system in hilly and mountainous regions. **(True)**
4. Fish farming is not suitable for coastal agro-climatic zones. **(False)** *(Coastal areas are ideal for aquaculture.)*
5. Dryland farming is best suited for regions with high annual rainfall. **(False)** *(Dryland farming is used in low-rainfall areas.)*
6. Climate change can impact the feasibility of farming systems in different agro-climatic zones. **(True)**
7. Soil type has no impact on the feasibility of farming systems. **(False)** *(Soil type determines crop suitability and fertility.)*
8. The availability of irrigation facilities can increase the feasibility of high-yield farming systems in dry regions. **(True)**
9. Shifting cultivation is commonly practiced in flood-prone areas. **(False)** *(Shifting cultivation is more common in hilly and forested regions.)*
10. Horticulture-based farming systems are more feasible in moderate-temperature agro-climatic zones. **(True)**
11. Rice-fish farming is commonly practiced in arid zones of Rajasthan **(False)**
12. Agroforestry is a suitable farming system in the Eastern Himalayan region. **(True)**

13. Trans-Gangetic Plains are ideal for silvipastoral systems. **(False)**
14. The Islands Region supports coastal integrated farming systems. **(True)**
15. Shifting cultivation is mainly practiced in the Western Plateau region. **(False)**
16. Protected cultivation is suitable for the Trans-Gangetic Zone. **(True)**
17. Gujarat Plains and Hills favor rice-based farming systems. **(False)**
18. The Lower Gangetic Plains are rich in water resources and support aquaculture. **(True)**
19. Apiculture is a feasible enterprise in the Western Himalayan Region. **(True)**
20. The Central Plateau is known for sericulture-based farming systems **(False)**

17

Commercial Farming-Based Livelihood Models by the NABARD

The National Bank for Agriculture and Rural Development (NABARD) plays a pivotal role in strengthening India's rural economy by facilitating credit flow, infrastructure development, and livelihood enhancement in agriculture and allied sectors. Among its many initiatives, the **promotion of commercial farming-based livelihood models** stands out as a dynamic and sustainable approach to empower rural communities—especially small and marginal farmers, tribal populations, and rural women. These livelihood models are **scientifically designed, economically viable, and socially inclusive**, aiming to generate income, create rural employment, and ensure food and nutritional security. They also align with national programs like the **Doubling Farmers' Income** mission, **National Livestock Mission**, **PM Matsya Sampada Yojana**, and the **Agri-Infra Fund**.

The National Bank for Agriculture and Rural Development (NABARD) in India has been instrumental in promoting sustainable agricultural practices and enhancing rural livelihoods through various commercial farming-based livelihood models. These models aim to empower farmers, increase production efficiency, and ensure better income through diversified agricultural practices. Here are some of the notable initiatives and models promoted by NABARD:

1. Dairy Farming as a Commercial Livelihood

Dairy farming is one of the most prominent models promoted by NABARD. Small dairy units with 2–10 crossbred cows or buffaloes are encouraged under schemes such as the **Dairy Entrepreneurship Development Scheme (DEDS)**. These units produce milk daily, generating regular income, and provide by-products like dung for compost or biogas. NABARD facilitates credit through banks and provides back-ended subsidies, especially for SHGs, youth, and women farmers. Technical training on fodder management, disease control, and scientific milking practices is also provided.

2. Poultry Farming: A Scalable Rural Enterprise

Poultry farming, particularly broiler and layer units, has shown great potential for income generation in rural and peri-urban areas. NABARD's models include **commercial broiler units of 1000–5000 birds**, backyard poultry for tribal families, and integrated hatcheries. Poultry requires lower initial investment and yields fast returns within 40–45 days for broilers. NABARD supports these models with refinance to banks and links entrepreneurs to markets via Farmer Producer Organizations (FPOs).

3. Goat Farming (Goatery) Models

Goat farming is especially suitable for resource-poor regions and is commonly referred to as the "poor man's cow." NABARD supports goat rearing units typically consisting of **10–20 does and 1–2 bucks**. High-yield breeds like Black Bengal, Sirohi, and Beetal are promoted depending on the region. These models are profitable due to low input costs and high market demand for goat meat. Linkages for fodder cultivation, veterinary support, and insurance are key components of this livelihood model.

4. Fisheries and Aquaculture

Aquaculture is a lucrative livelihood model in water-abundant areas like Eastern India, the Northeast, and coastal states. NABARD promotes **pond-based fish farming**, integrated rice-fish systems, and shrimp/prawn culture in saline regions. Beneficiaries are supported with funds for pond construction, fingerling stocking, feed, and equipment. These models generate income within 6–8 months and provide nutritional security. NABARD also promotes the formation of **Fish Farmer Producer Organizations (FFPOs)** to strengthen market access.

5. Horticulture-Based Farming Models

With rising demand for fruits and vegetables, horticulture offers profitable commercial opportunities. NABARD supports **high-density plantation models** involving mango, banana, pomegranate, and guava cultivation. These are often combined with intercropping, drip irrigation, and vermicomposting. Beneficiaries are encouraged to set up **nurseries, pack houses, and cold storage units** for value addition and shelf-life extension. The returns are substantial, especially after the gestation period of 2–3 years, and subsidies are provided under the Mission for Integrated Development of Horticulture (MIDH).

6. Mushroom Cultivation: High Returns in Small Spaces

Mushroom farming, particularly **oyster, milky, and button varieties**, has been widely adopted as a livelihood model, especially among landless women and SHGs. It requires low space and investment, making it ideal for backyard or shed-level production. NABARD offers capacity building and financial support through microfinance institutions. With a crop cycle of just 45–60 days, farmers can harvest mushrooms multiple times a year and generate steady income.

7. Sericulture and Silk Production

Sericulture is a highly labor-intensive but profitable enterprise promoted by NABARD in states like Karnataka, Jharkhand, and West Bengal. The livelihood model includes **mulberry plantation, silkworm rearing, and cocoon reeling units**. NABARD assists with infrastructure like rearing houses, irrigation, and reeling equipment. These models provide employment to the entire household, and the silk produced fetches high value in both domestic and international markets. They also integrate well with agroforestry and provide long-term benefits.

8. Agroforestry and Bamboo-Based Models

To promote climate-resilient livelihoods, NABARD encourages farmers to adopt **agroforestry models** integrating tree crops like teak, bamboo, poplar, or eucalyptus with seasonal crops. These long-term investments offer timber, fodder, and non-timber forest products (NTFPs), diversifying farmers' income. Bamboo in particular is promoted for use in construction, handicrafts, and furniture. NABARD supports this model through interest subvention, marketing assistance, and technical guidance via Krishi Vigyan Kendras (KVKs).

9. Apiculture (Beekeeping)

Apiculture, or beekeeping, is another promising model offering both **direct income from honey and wax**, and **indirect benefits like crop pollination**. Farmers are trained to maintain bee colonies in modern hives and extract honey hygienically. Typically, a unit consists of 50–100 bee boxes, and honey can be sold locally or processed for value addition. NABARD provides financial and technical support for setting up these units and linking them with processing and branding enterprises under FPOs.

10. Integrated Farming Systems (IFS):

The most holistic model promoted by NABARD is the **Integrated Farming System**, where a combination of crop production, livestock, aquaculture, horticulture, and agroforestry are practiced on the same landholding. This model maximizes resource use efficiency, reduces risk, and ensures year-round income. It is customized to suit local agro-climatic conditions. NABARD promotes IFS models in convergence with schemes like MGNREGA, National Rural Livelihood Mission (NRLM), and State Agriculture Missions.

- NABARD encourages integrated farming systems that combine crop production, livestock, aquaculture, and agroforestry, enhancing resource use efficiency and farm income.
- IFS models promote sustainability and reduced risk for farmers by diversifying income sources.

11. Organic Farming Initiatives

- NABARD promotes organic farming methods which reduce dependency on chemical inputs, enhance soil health, and increase demand for organic products.
- Training programs and certifications for organic farming help farmers tap into premium markets.

12. Agri-Entrepreneurship Development

- NABARD supports agri-entrepreneurship by facilitating skill development programs and financial assistance for startups in agribusiness.
- This includes processing, value addition, and marketing of agricultural produce, thereby increasing farmers' income.

13. Climate-Smart Agriculture

- NABARD promotes climate-smart agricultural practices to enhance resilience to climate change while ensuring sustainable production.
- Initiatives include water conservation, soil health management, and sustainable land-use practices.

14. Micro-irrigation and Water Conservation

- Investments in micro-irrigation technologies, such as drip and sprinkler systems, are promoted to enhance water use efficiency in farming.
- NABARD provides financial support and technical assistance to farmers adopting these practices.

15. Skill Development and Training Programs

- NABARD develops and implements various training programs aimed at equipping farmers with modern agricultural techniques and market information.
- Focus areas include pest management, post-harvest handling, and market linkages.

16. Digital Agriculture Initiatives

- NABARD has been promoting the use of technology in agriculture, including mobile applications for advisory services, market information, and e-commerce platforms for selling produce.
- Digital initiatives help in bridging the information gap for farmers.

17. Value Chain Development

- NABARD supports value chain development for various agricultural products, ensuring that farmers receive a fair price and that there is a systematic approach to production, processing, and marketing.

These livelihood models focus on enhancing productivity, ensuring sustainable practices, and increasing rural incomes. NABARD plays a crucial role in providing financial support, capacity-building, and facilitating partnerships between stakeholders, including government agencies, private sector players, and civil society organizations. Through these initiatives, NABARD aims to foster a more resilient and profitable agricultural sector in India, contributing to rural development and poverty alleviation.

QUESTION BANK

Short Answer Questions

1. What is NABARD, and what is its role in promoting commercial farming?
2. Name two commercial farming-based livelihood models supported by NABARD.
3. How does NABARD support agripreneurs in commercial farming?
4. What are the key components of NABARD's integrated farming system models?
5. Name a major NABARD scheme that supports commercial dairy farming.
6. How does NABARD help small and marginal farmers adopt commercial farming?
7. What is the role of Farmer Producer Organizations (FPOs) in NABARD's commercial farming model?
8. How does NABARD promote value chain development in commercial farming?
9. What are the major challenges in implementing NABARD-supported commercial farming models?
10. How does NABARD assist in market linkage for commercial farmers?

Multiple Choice Questions (MCQs)

1. **What is the primary goal of NABARD's commercial farming-based livelihood models?**
   a) Providing employment in urban areas
   b) Enhancing agricultural productivity and rural incomes
   c) Increasing government tax collection
   d) Reducing the use of fertilizers

2. **Which of the following commercial farming enterprises is supported by NABARD?**
   a) `IT services b) Dairy farming
   c) Automobile manufacturing d) Hotel business

3. **Which NABARD initiative supports rural entrepreneurs in commercial farming?**
   a) MGNREGA
   b) Self Help Group-Bank Linkage Program
   c) Pradhan Mantri Awas Yojana
   d) National Highway Development Program

4. **Which of the following best describes an integrated farming system promoted by NABARD?**
   a) A single crop production model
   b) Combining crops, livestock, and allied activities for sustainability
   c) Complete reliance on chemical fertilizers
   d) Farming without irrigation

5. **What is the primary function of NABARD's Farmer Producer Organizations (FPOs)?**
   a) To regulate food prices
   b) To promote collective marketing and enhance bargaining power
   c) To control water supply for irrigation
   d) To manufacture fertilizers

6. **Which of the following is a key factor for the success of NABARD's commercial farming models?**
   a) Land fragmentation
   b) Market linkages and financial support
   c) Ban on agricultural exports
   d) Lack of modern technology

7. **NABARD provides financial assistance to commercial farming enterprises through**
   a) Direct lending only
   b) Refinancing banks and financial institutions
   c) Social media campaigns
   d) Foreign investment schemes

8. **Which NABARD scheme provides subsidies for dairy and poultry farming?**
   a) Kisan Credit Card Scheme
   b) Dairy Entrepreneurship Development Scheme (DEDS)
   c) Rashtriya Swasthya Bima Yojana
   d) Jan Dhan Yojana

9. **Why is crop diversification encouraged in NABARD's commercial farming models?**
   a) To reduce soil fertility
   b) To minimize risks and improve income stability
   c) To increase dependency on imported fertilizers
   d) To restrict farmers to one type of crop

10. **What is the major benefit of NABARD's Warehouse Infrastructure Fund for farmers?**
    a) Allows farmers to store produce and sell at better prices
    b) Increases import of foreign grains
    c) Reduces the use of organic farming practices
    d) Prevents farmers from accessing bank loans

True/False Questions

1. NABARD supports commercial farming models to improve rural employment opportunities. **(True)**
2. NABARD discourages the use of technology in commercial farming. **(False)** *(It promotes modern technology for better efficiency.)*
3. The Dairy Entrepreneurship Development Scheme (DEDS) is an initiative under NABARD. **(True)**
4. NABARD's financing is only available to large-scale farmers. **(False)** *(It supports small, marginal, and large farmers.)*
5. Integrated farming systems promoted by NABARD include livestock, crops, and fisheries. **(True)**
6. NABARD does not provide support for organic farming. **(False)** *(It encourages organic farming practices.)*
7. NABARD's Farmer Producer Organizations (FPOs) help small farmers access better markets. **(True)**
8. NABARD only focuses on credit and does not support infrastructure development. **(False)** *(It also funds rural infrastructure projects.)*
9. Value chain development is a key component of NABARD-supported commercial farming models. **(True)**
10. NABARD supports agri-businesses and processing units along with commercial farming. **(True)**

18

Commercial Farming-Based Livelihood Models by the ICAR

The Indian Council of Agricultural Research (ICAR) has developed various commercial farming-based livelihood models to enhance the income of farmers, promote sustainable agricultural practices, and improve the overall socio-economic conditions in rural areas. While specific models may vary by region and crop, here are some key features and themes observed in ICAR's initiatives:

1. Integrated Farming System (IFS) Models

The **Indian Council of Agricultural Research (ICAR)** has developed a wide range of **Integrated Farming System (IFS) models** tailored to various agro-climatic zones of India. These models aim to enhance **farm productivity**, ensure **sustainable livelihood**, and improve **resource use efficiency** by integrating different components like crops, livestock, poultry, fishery, horticulture, agroforestry, etc.

Key Objectives of ICAR's IFS Models

- Enhance income and employment from small and marginal holdings.
- Ensure year-round productivity and food security.
- Recycle resources within the farm to minimize external input dependency.
- Promote sustainability through organic integration and ecological balance.

Tables: Examples of ICAR-Developed IFS Models

Agro-Climatic Zone	Institute / KVK	Major Components Integrated	Annual Income (Approx.)
Eastern India (Bihar, Odisha)	ICAR-RCER, Patna	Rice–Vegetable–Fish–Goat	□1.25 to □1.50 lakh/ha
North-Western Plains (Punjab)	ICAR-IARI & KVK Ludhiana	Crop–Dairy–Biogas–Fodder	□2.00 lakh/ha
Eastern Plateau (Jharkhand)	BAU, Ranchi & ICAR-KVKs	Crop–Poultry–Fish–Horticulture	□1.10 to □1.35 lakh/ha

Western Dry Region (Rajasthan)	ICAR-CAZRI, Jodhpur	Pearl millet–Legume–Goat–Agroforestry	□90,000–□1.2 lakh/ha
Coastal Region (Andhra Pradesh)	ANGRAU, ICAR-CIBA	Paddy–Fish–Poultry–Dairy	□1.50–□2.00 lakh/ha
North Eastern Hills (Nagaland)	ICAR-RCNEH, Jharnapani	Terrace rice–Pig–Poultry–Agroforestry	□1.00–□1.3lakh/ha
Central India (MP, Chhattisgarh)	IGKV, Raipur & ICAR-IIFSR	Rice–Pulse–Vegetable–Dairy–Poultry	□1.50 lakh/ha
Hill & Mountain Region (Uttarakhand)	VPKAS, Almora	Cereal–Horticulture–Goat–Bee–Agroforestry	□90,000 to □1.1 lakh/ha

Typical Components of ICAR IFS Models

1. **Crops**: Cereals, pulses, oilseeds, vegetables.
2. **Livestock**: Dairy, goatry, piggery.
3. **Poultry/Duckry**: Backyard poultry, Vanaraja, Gramapriya.
4. **Fisheries**: Composite fish culture, biofloc (new tech).
5. **Horticulture**: Fruits (banana, papaya, guava), vegetables.
6. **Mushroom**: Oyster, button cultivation.
7. **Beekeeping**: Honey and wax production.
8. **Agroforestry**: MPTs (Multi-purpose trees), boundary plantations.
9. **Biogas/Composting**: Waste recycling and energy.

Impact of ICAR's IFS Models

- Income enhanced by **2 to 3 times** in many regions.
- Improved **nutritional security** of farm families.
- Increased **employment generation** (200–300 man-days/year/ha).
- Significant **reduction in input costs** through internal recycling.
- Promoted **women participation** and youth engagement in agriculture.

Research & Coordination Institutes Involved

- ICAR-IIFSR (Modipuram) – Nodal agency for IFS research.
- ICAR-RCER (Patna) – Eastern India focused IFS models.
- ICAR-CAZRI (Jodhpur) – Dryland and arid zone models.
- ICAR-ATARI Zones – Technology demonstrations via KVKs.

ICAR's **Integrated Farming System (IFS)** models offer a **holistic, sustainable, and income-enhancing approach** to agriculture, especially for **small and marginal farmers**. They are pivotal in achieving the goals of **Doubling Farmers' Income, resource sustainability**, and **rural employment generation**.

2. Dairy-Based Livelihood Model

A **Dairy-Based Livelihood Model** is a farming enterprise centered on milk production, processing, and marketing, integrated with fodder cultivation, manure management, and often other supportive components like biogas, composting, and value-added dairy products. It provides a **sustainable income source** for rural households, especially **small and marginal farmers**.

Objectives of the Dairy Model

- Generate regular daily income for farm families.
- Improve nutritional security (milk and milk products).
- Utilize crop residues and biomass effectively.
- Promote women's participation and employment.
- Provide integrated recycling through dung for compost or biogas.

Typical Structure of a Dairy-Based Livelihood Model

Component	Description
Dairy Unit	2–10 cows/buffaloes (high-yielding or indigenous breeds)
Fodder Cultivation	Napier, Berseem, Maize, Lucerne grown on-farm
Crop Integration	Crop residues (wheat/paddy straw) used as feed
Manure Management	Dung used for composting or biogas generation
Milk Processing	Small-scale production of curd, paneer, ghee (value-added products)
Marketing	Milk collection through cooperatives/self-help groups
Support Services	Veterinary care, AI services, training via ICAR/KVK

Regional Examples (Promoted by ICAR)

Region	Model Features	Support Institute
Punjab, Haryana	5–10 crossbred cows, fodder block system, biogas unit	ICAR-NDRI, Karnal
West Bengal, Odisha	2–5 indigenous/crossbred cows, milk cooperatives	ICAR-Eastern Region KVKs
Tamil Nadu	Dairy + Azolla feeding, backyard milk processing	TANUVAS, KVKs
Gujarat	Buffalo-based milk production + value addition (ghee, sweets)	NDDB, ICAR-KVKs

Income Generation Potential

- **Smallholder (2–3 animals)**: □50,000–□80,000/year
- **Medium unit (5–8 animals)**: □1.2–□2.0 lakh/year
- **Large unit (10+ animals)**: □3.0 lakh and above/year *(Income depends on milk yield, breed, feed quality, and market access)*

Innovations Supported by ICAR

- Azolla as cattle feed supplement.
- Use of mineral mixture and probiotics for better milk yield.
- Low-cost cattle shed design.
- Use of ICT tools for animal health and estrus monitoring.

Gender and Livelihood Impact

- Promotes **women's empowerment** through household-level dairying.
- **Self-help groups (SHGs)** are often engaged in milk marketing and value addition.
- Creates **daily cash flow**, improving rural financial stability.

Benefits of the Dairy Model

- Regular income with minimal land.
- Nutritional and food security.
- Efficient recycling of crop waste and dung.
- High employment potential.
- Climate-resilient and location-flexible.

3. Poultry-Based Farming Models

Poultry-based farming models involve rearing birds such as chickens, ducks, quails, or turkeys primarily for **egg and meat production**. These models can be implemented as **backyard systems** or **commercial enterprises**, offering **quick returns**, **low investment**, and **high income potential**, especially for **small and marginal farmers** and **landless families**.

Objectives of Poultry-Based Models

- Enhance rural income and employment.
- Ensure household-level nutritional security (eggs & meat).
- Engage women and youth in livelihood generation.
- Integrate with other farm components for resource recycling.

Components of a Poultry-Based Livelihood Model

Component	Description
Bird Type	Broilers (meat), Layers (eggs), Dual-purpose (e.g., Vanaraja, Gramapriya)
Housing	Low-cost poultry sheds or mobile poultry units
Feed & Nutrition	Maize, soya, broken rice, kitchen waste; supplemented with minerals

Health Management	Vaccination, deworming, and biosecurity measures
Market Linkages	SHGs, cooperatives, local haats or online platforms
Integration	Can be combined with dairy, fishery, vermicomposting, crop farming, etc.

Types of Poultry Farming Models

Model	Details
Backyard Poultry Model	10–50 birds, low-cost, low-maintenance, ideal for rural households
Semi-Commercial Model	200–500 birds, requires housing, feed & market planning
Commercial Broiler Model	1000+ birds, raised for meat, 35–45 days cycle
Commercial Layer Model	1000+ birds for egg production, 18–20 months cycle
Duckery cum Fishery Model	Ducks feed on pond insects; droppings fertilize pond for fish production
Quail or Japanese Quail Unit	High-density bird rearing in cages, urban and peri-urban areas

ICAR-Recommended Poultry Breeds for Rural Models

Breed	Purpose	Benefits
Vanaraja	Dual (egg & meat)	Hardy, scavenging, colorful plumage
Gramapriya	Egg production	Early layer, disease-resistant
Kuroiler	Meat & egg	Fast-growing, adaptable to rural conditions
Kadaknath	Specialty meat	High market value, native to Madhya Pradesh
Chabro	Broiler	Fast-growing, suited for semi-intensive model

Income Potential

Model Type	Birds	Cycle Duration	Estimated Annual Income
Backyard Poultry	20–50	Ongoing	□20,000–□50,000
Semi-commercial Broiler	500	6–8 cycles/year	□1.5–□2.5 lakh
Commercial Layer	1000	12–18 months	□2.5–□4.0 lakh

Livelihood & Social Impact

- Empowers women through SHGs and home-based units.
- Provides food and cash security.
- Engages youth in agripreneurship and enterprise models.
- Suitable for landless and marginal farmers.

Integration Opportunities

- **Crop–Poultry–Fishery**: Poultry droppings fertilize fish ponds.
- **Dairy–Poultry**: Milk by-products used in feed; compost from bedding.
- **Biogas–Poultry**: Poultry litter used in biogas or vermicomposting.

Support Organizations

- **ICAR-CARI** (Central Avian Research Institute)
- **KVKs and ATMA**: Training and extension
- **NABARD & State Livestock Missions**: Subsidies and financial support
- **NGOs/CSR**: Project-based model promotion

4. Goatry and Small Ruminant-Based Models

Goatry and small ruminant-based farming models focus on the rearing of goats and sometimes sheep for **meat, milk, skin, and manure**. These models are particularly suitable for **resource-poor farmers**, **marginal lands**, and **drought-prone areas**, offering **low investment**, **minimal input**, and **high returns**, especially under **semi-intensive or backyard systems**.

Objectives of Goatry-Based Models

- Promote livelihood generation in rural, tribal, and arid zones.
- Enhance meat and milk availability.
- Utilize degraded or marginal land and non-crop biomass.
- Empower landless farmers, women, and youth.
- Improve economic resilience through diversified income.

Components of Goatry-Based Livelihood Model

Component	Description
Breed Selection	Native (Black Bengal, Barbari, Osmanabadi), Crossbreds, or Exotics (Boer, Sirohi)
Housing	Low-cost shed with slatted floor (raised platform)
Feeding System	Browse plants, tree leaves, kitchen waste, concentrated feed, crop residues
Health Care	Vaccination (PPR, FMD, HS), deworming, veterinary care
Reproduction Management	Strategic kidding, use of superior bucks or AI
Marketing	Live animals, meat, milk, and manure sales
Integration Options	Crop-goat, poultry-goat, goat-fish, goat-forestry

Popular Models Based on Region & Enterprise Type

Model	Region	Key Features
Backyard Goatry	West Bengal, Bihar, Odisha	2–10 goats, low-input, stall + free grazing
Semi-Intensive Model	Rajasthan, Gujarat	20–50 goats, mixed feeding, commercial meat sale
Intensive Model	Urban fringe zones	50–200 goats, stall-fed, commercial dairy & meat unit
Sheep-Goat Mixed Unit	Southern plateau (TN, AP)	Grazing-based system with dual species

Recommended Goat Breeds in India

Breed	Region	Utility
Black Bengal	Eastern India	Meat, early maturity
Barbari	Northern India	Dual-purpose
Osmanabadi	Deccan Plateau	Meat & milk
Sirohi	Rajasthan	Large body, meat
Malabari	Kerala	Milk and adaptability

Estimated Income from Goatry

Scale	No. of Goats	Cycle/Year	Annual Income (□)
Backyard	5–10	1–2 kiddings	□30,000–□50,000
Semi-commercial	25–50	2 kiddings	□1.0–□2.0 lakh
Intensive	100–200	2 kiddings	□3.0–□6.0 lakh

Integration Possibilities

- **Crop–Goat Model**: Goats fed on crop residues; dung used in fields.
- **Goat–Poultry Model**: Shared housing, complementary feed use.
- **Goat–Agroforestry Model**: Browsing on multipurpose trees/shrubs.
- **Goat–Fish Model**: Goat waste fertilizes fish pond.

Livelihood and Gender Empowerment

- Ideal for **tribal women SHGs** and **landless households**.
- Quick turnover through meat sales.
- Facilitates **micro-enterprise development** in hilly and dryland regions.

Support Institutions

- **ICAR-CIRG (Central Institute for Research on Goats), Mathura**
- **KVKs** (Krishi Vigyan Kendras) in goat-dominant zones
- **NABARD** for model bankable projects and subsidies
- **State Livestock Missions & NRLM** for SHG/Cluster-based promotion

5. Fish-Based Livelihood Models

Fish-based livelihood models refer to **aquaculture and capture fisheries systems** developed for income, employment, and nutritional support in **rural, peri-urban, and coastal regions**. These models contribute significantly to **food security**, **sustainable livelihoods**, and **rural economy diversification**, especially in **water-rich agro-climatic zones** of India.

Objectives of Fish-Based Models

- Utilize ponds, tanks, canals, paddy fields, and wetlands for income generation.
- Improve protein availability at the household level.
- Provide employment to rural youth and women.
- Promote integrated farming by linking fishery with livestock, crops, and horticulture.
- Support sustainable management of aquatic resources.

Components of a Fish-Based Livelihood Model

Component	Description
Water Resource	Ponds, tanks, ditches, wetlands, rice fields, cages, pens, RAS systems
Fish Species	Rohu, Catla, Mrigal, Tilapia, Pangasius, Common Carp, Magur, Singhi, Pabda
Culture System	Mono-culture, polyculture, integrated systems (e.g., fish–duck, fish–paddy)
Feed Management	Commercial pellets, farm-based feed (rice bran, mustard cake, cow dung, etc.)
Health & Water Quality	Regular monitoring of pH, DO, turbidity, ammonia; vaccination and prophylactics
Marketing & Processing	Direct sale, cooperatives, cold chain, value-added products

Types of Fish-Based Livelihood Models

Model	Description
Pond-Based Aquaculture	Freshwater fish culture in household/community ponds
Paddy–Fish Integration	Seasonal rice fields with fish culture, especially in Eastern & NE India
Cage/Pen Culture in Reservoirs	Floating cages in reservoirs/lakes for fish fattening
Fish–Duck Integration	Ducks fertilize water; fish feed on plankton and duck droppings
Fish–Pig Integration	Pig waste used to enhance pond productivity
Fish–Horticulture System	Bunds of ponds used for vegetables/fruit trees
Recirculatory Aquaculture System (RAS)	Water-saving, indoor, high-density fish culture system
Brackish Water Aquaculture	Coastal aquaculture (e.g., shrimp, crab) for saline water regions

Common Fish Species by Model

Species	Suited For
Rohu, Catla	Polyculture in freshwater ponds
Tilapia	Backyard tanks, RAS, cage culture
Magur, Singhi	Tank-based or biofloc culture
Pangasius	High-density commercial farming
Shrimp (P. vannamei)	Brackish water aquaculture

Income Potential (Indicative)

Model	Area	Cycle	Estimated Annual Income (□)
Backyard Pond Unit	0.05 ha	1–2 cycles	□40,000–□60,000
Semi-Commercial Pond	0.2 ha	1 cycle	□1.5–2.5 lakh
Paddy-Fish Culture	1 acre	Kharif season	□30,000–□50,000 (in addition)
Cage Culture	1 cage	3–4 months	□50,000–□1.0 lakh/cage
Shrimp Farming	1 ha	2 crops	□4.0–6.0 lakh

Integration Opportunities

- **Duck–Fish**: Ducks fertilize water and control insects.
- **Crop–Fish–Livestock**: Effective nutrient recycling and year-round inc'ome.
- **Horticulture–Fish**: Pond bunds used for vegetables or fruit crops.
- **Vermicompost–Fish**: Organic waste management and enhanced pond fertility.

Livelihood and Social Impact

- Boosts incomes for small/marginal farmers, tribal families, and women.
- Creates local jobs in feed supply, net making, hatchery, and processing.
- Encourages entrepreneurship through fish seed production and sale.

Support Institutions

- **ICAR-CIFA** (Central Institute of Freshwater Aquaculture), Bhubaneswar
- **NFDB** (National Fisheries Development Board)
- **State Fisheries Departments**
- **NABARD & PMMSY** (for financing and subsidies)
- **KVKs & NGOs** for capacity building

6. Commercial Horticulture-Based Models

Commercial horticulture-based models refer to **organized, market-oriented production** of **fruits, vegetables, flowers, spices, plantation crops, and**

medicinal & aromatic plants. These models are designed to **maximize returns per unit area**, generate **employment**, ensure **nutritional security**, and strengthen the **agriculture value chain**.

Objectives of Horticulture-Based Models

- Enhance farmers' income through high-value crops.
- Utilize land, water, and climate resources optimally.
- Promote diversification from traditional cereal-based farming.
- Encourage entrepreneurship in post-harvest processing and export.
- Generate employment, especially for youth and women.

Major Components of Commercial Horticulture Models

Component	Description
Crop Selection	High-value crops like mango, banana, guava, onion, tomato, marigold, ginger etc.
Input Management	Use of quality planting materials, drip irrigation, fertigation, mulching
Protected Cultivation	Polyhouse, net house, greenhouse structures for off-season and export crops
Integrated Nutrient & Pest Management	Use of bio-inputs, IPM/INM techniques
Post-Harvest Handling	Grading, cold storage, packaging, processing
Market Linkages	Farmer Producer Organizations (FPOs), contract farming, e-marketing

Types of Commercial Horticulture Models

Model	Key Features
Fruit-Based Orchard Systems	Mango, guava, banana, papaya, citrus; long-term income with intercropping
Vegetable Production Units	Tomato, chili, onion, okra, cole crops; suitable for peri-urban markets
Floriculture Models	Marigold, rose, jasmine, gladiolus; grown in open fields or protected units
Spice & Aromatic Models	Ginger, turmeric, lemongrass, tulsi; value-added products & export
Protected Horticulture Units	Polyhouse/net house-based farming for capsicum, cucumber, strawberry, etc.
High-Density & Ultra-High-Density Orchards	For apple, guava, mango, citrus in small land holdings
Horticulture + Beekeeping	Enhances pollination and honey production
Agri-Tourism with Horticulture	Combining farm visits with orchard activities for income diversification

Economic Potential (Indicative)

Model	Area (per unit)	Expected Net Annual Income (□)
Vegetable Cultivation	0.5 acre	□80,000 – □1,50,000
Fruit Orchard (e.g., Guava)	1 acre	□1.5 – □3 lakh
Polyhouse Cultivation	1000 sq.m.	□2 – □4 lakh
Flower Farming (Marigold)	1 acre	□1 – □2 lakh
Ginger/Turmeric Farming	1 acre	□1.5 – □3 lakh

Suitable Agro-Climatic Zones & Crops

Zone	Suitable Crops
Eastern India	Banana, papaya, litchi, jackfruit, ginger, turmeric
Western India	Pomegranate, ber, citrus, onion
Northern Hills	Apple, pear, plum, walnut, medicinal herbs
Southern India	Coconut, areca nut, black pepper, curry leaves, banana
Indo-Gangetic Plains	Mango, guava, flowers, cucurbits, mustard for intercropping

Value Addition Opportunities

- **Fruit pulping, drying, pickles, jams**
- **Spice powders, essential oils**
- **Cut flowers, bouquets, garlands**
- **Cold chain and packaging**
- **Export-ready certification (GAP, organic)**

Institutional Support

- **ICAR-IIHR, ICAR-CISH, ICAR-IIVR** for technologies & varieties
- **MIDH, NHM, SFAC, NABARD** for schemes, credits, and subsidies
- **FPOs and SHGs** for collective marketing
- **State Horticulture Departments** for training and market support

Integrated Opportunities

- **Horti–Bee–Mushroom Models**
- **Horti–Dairy or Horti–Fish Integration**
- **Horticulture–Poultry on bunds or inter-row areas**

7. Mushroom Production Model

The **Mushroom Production Model** refers to the systematic cultivation of edible fungi such as **button mushroom, oyster mushroom, milky mushroom, and paddy straw mushroom** under controlled or semi-controlled conditions. This model offers a **sustainable, low-space, low-input livelihood option**, especially for landless, marginal, and smallholder farmers.

Objectives

- Utilize agricultural residues (wheat straw, paddy straw, sawdust, etc.).
- Generate income in small spaces, including indoors or backyards.
- Promote nutrition through protein-rich fungi.
- Empower rural youth, women, and self-help groups (SHGs).
- Enhance circular economy by recycling biomass.

Types of Mushrooms Cultivated in India

Type of Mushroom	Scientific Name	Best Season
Button Mushroom	*Agaricus bisporus*	Winter (15–20°C)
Oyster Mushroom	*Pleurotus spp.*	Year-round (20–30°C)
Paddy Straw Mushroom	*Volvariella volvacea*	Summer (30–35°C)
Milky Mushroom	*Calocybe indica*	Summer (25–35°C)
Shiitake Mushroom	*Lentinula edodes*	Temperate hills

Components of the Model

Component	Details
Spawn production/ availability	Certified spawn from ICAR institutes or state agencies
Substrate management	Paddy straw, wheat straw, sugarcane bagasse, sawdust
Cultivation unit	Sheds, rooms, bamboo/poly tunnels with humidifiers and ventilation
Harvesting & grading	Manual harvesting at maturity; sorting by size and quality
Storage & processing	Short shelf life; drying, pickling, powdering as value addition
Marketing	Direct sale, local markets, tie-up with aggregators or restaurants

Economics (Indicative for Oyster Mushroom)

Particulars	Quantity/Cost (Approx.)
Area Required	100 sq. ft (1 small room)
Initial Setup Cost	□10,000 – □15,000
Substrate (wheat/paddy straw)	□2,000
Spawn	□3,000
Labor & Misc.	□2,000
Yield per cycle	100–120 kg
Market Price	□120–□150/kg
Net Return (1 cycle)	□10,000 – □12,000 (in 30–40 days)

Institutional Support

- **ICAR-DMR (Directorate of Mushroom Research), Solan (HP)**
- **KVKs and ATMA projects**

- **State Horticulture Missions**
- **NABARD & MSME schemes** for training and subsidy
- **Agri-clinics** and entrepreneurship incubation centers

8. Integration with Other Systems

Enterprise	Mode of Integration
Vermicomposting	Mushroom spent substrate used for composting
Horticulture/IFS	Spent compost improves soil organic matter
Poultry/Fishery	Income diversification and reuse of organic waste
Agro-Tourism	Demonstration units attract educational and tourism interest

Advantages

- Low investment and space
- High income per unit area
- Suitable for women and marginalized groups
- Year-round income potential
- Enhances nutritional security

10. Beekeeping (Apiculture) Models

Beekeeping or **apiculture** is the scientific management of honeybee colonies for the production of **honey, beeswax, royal jelly, pollen, and propolis**, along with promoting **pollination services** in agricultural and horticultural crops. It is a **low-investment, high-return enterprise** that can be integrated with farming systems to **enhance productivity and income**.

Objectives

- Generate income from honey and other bee products.
- Enhance crop yield through pollination services.
- Provide rural employment, especially for small/marginal farmers.
- Support biodiversity and ecological balance.
- Strengthen allied value chains in food and cosmetic industries.

Major Species Used in India

Bee Species	Scientific Name	Common Habitat
Indian Honeybee	*Apis cerana indica*	Hilly and plains, managed in boxes
European/Italian Honeybee	*Apis mellifera*	Plains and commercial beekeeping units
Rock Bee	*Apis dorsata*	Forests and rocky cliffs (wild, non-box)
Little Bee	*Apis florea*	Bushy areas, not preferred for domestication
Stingless Bee	*Trigona spp.*	South India, for medicinal honey

Beekeeping Model Components

Component	Details
Bee Boxes/Colonies	Langstroth boxes with colonies (5–20 units)
Location & Shelter	Shaded, cool, and pesticide-free zone; under trees or sheds
Floral Resources	Plantation of bee-friendly crops (mustard, sunflower, citrus, guava, eucalyptus)
Feeding & Watering	Sugar syrup and clean water during lean seasons
Seasonal Management	Shifting hives near crops, honey extraction twice a year
Product Harvesting	Honey, wax, pollen, royal jelly (depending on species)
Processing & Packaging	Simple filtration, bottling, branding, and marketing

Economics (Indicative)

Particulars	Unit (10 colonies)
Initial Investment	□25,000 – □35,000
Annual Honey Yield	200–300 kg
Market Price of Honey	□150 – □300/kg
Gross Annual Return	□45,000 – □90,000
Net Income (after expenses)	□30,000 – □60,000

Integration with Other Farming Enterprises

Integrated System	Benefit
Horticulture + Beekeeping	Enhanced pollination (fruit set/yield up by 20–30%)
Agroforestry + Apiculture	Trees provide shelter and year-round nectar
Crop Fields (Mustard, Sunflower)	Seasonal relocation of boxes for pollination and honey
IFS & Agro-Tourism	Adds educational and aesthetic value for visitors

Institutional Support

- **ICAR-Central Bee Research & Training Institute (CBRTI), Pune**
- **KVKs and State Apiculture Missions**
- **National Bee Board (NBB) under MIDH**
- **NABARD/NSDC** support for training and credit
- **Beekeeping cooperatives, SHGs, FPOs** for collective production and marketing

Value-Added Products

Product	Uses
Honey	Natural sweetener, medicine
Beeswax	Candles, cosmetics, polish
Royal Jelly	Health supplements, nutraceuticals
Pollen/Propolis	Immune boosters, pharma industry

Some other models aim to transform traditional agricultural practices these are given below:

1. Diversified Cropping Systems

- Encouraging farmers to adopt diversified models that combine traditional crops with high-value crops, such as fruits, vegetables, and cash crops, to ensure profitability and reduce risk.

2. Value Chain Development

- Focusing on building efficient value chains that connect farmers to markets. This includes post-harvest management, processing, packaging, and distribution, aimed at enhancing the marketability of agricultural products.

3. Integrated Farming Systems (IFS)

- Promoting IFS that combine crop production with livestock, aquaculture, and agroforestry, which helps in optimizing resource use, increasing productivity, and generating multiple income streams for farm families.

4. Agroecological Practices

- Encouraging sustainable practices such as organic farming, conservation agriculture, and agroecology, which help in maintaining soil health, reducing dependency on chemical inputs, and increasing resilience to climate change.

5. Entrepreneurship and Skill Development

- Providing training programs for farmers to develop skills in entrepreneurship, marketing, and agri-tech, enabling them to establish their own agribusiness ventures or cooperatives.

6. Cluster-Based Approach

- Implementing a cluster-based approach to agriculture that involves groups of farmers working together to share resources, knowledge, and access to markets, fostering collaboration and improving collective bargaining power.

7. Use of Technology and Innovation

- Leveraging technology, including precision farming, remote sensing, and mobile applications for market access and weather forecasting, to improve productivity and decision-making.

8. Financial Inclusion:

- Facilitating access to credit and insurance products for farmers to ensure they have the financial means to invest in better practices and technologies.

9. Public-Private Partnerships

- Encouraging partnerships between government agencies, research institutions, NGOs, and the private sector to create synergies and develop comprehensive models for commercial farming.

10. Community Engagement

- Involving local communities in the planning and implementation of agricultural initiatives to ensure that the models are culturally appropriate and meet the specific needs of the farmers.

These models aim to transform traditional agricultural practices into more profitable and sustainable commercial ventures, helping to uplift farmers' livelihoods and contribute to rural development in India.

QUESTION BANK

Short Answer Questions

1. What is ICAR, and how does it contribute to commercial farming?
2. Name two commercial farming-based livelihood models developed by ICAR.
3. How does ICAR support integrated farming systems for livelihood enhancement?
4. What role does ICAR play in promoting precision agriculture?
5. Name one ICAR initiative that supports dairy-based livelihood models.
6. How does ICAR contribute to sustainable commercial aquaculture?
7. What is the significance of agri-business incubators supported by ICAR?
8. How do ICAR's farming system research models help small and marginal farmers?
9. What is the role of ICAR's Krishi Vigyan Kendras (KVKs) in commercial farming?
10. How does ICAR promote value addition and agro-processing in farming?

Multiple Choice Questions (MCQs)

1. **Which of the following is a primary focus of ICAR in commercial farming?**
   a) Promoting IT services
   b) Enhancing agricultural productivity and farmer incomes
   c) Expanding urban infrastructure
   d) Importing food grains

2. **Which ICAR initiative supports commercial fish farming?**
   a) Rashtriya Krishi Vikas Yojana
   b) National Innovations in Climate Resilient Agriculture (NICRA)
   c) Indian Institute of Fisheries Research (IIFR) programs
   d) National Solar Mission

3. **Which ICAR institute specializes in dairy farming research?**
   a) Indian Institute of Pulses Research
   b) National Dairy Research Institute (NDRI)
   c) Indian Institute of Horticultural Research
   d) Indian Institute of Sugarcane Research

4. **Which farming model is promoted by ICAR for sustainable commercial vegetable farming?**
   a) Shifting cultivation
   b) Protected cultivation under polyhouses
   c) Slash-and-burn agriculture
   d) Pastoral nomadism

5. **ICAR supports the integration of which farming components for sustainable livelihood?**
   a) Agriculture, livestock, and fisheries
   b) Only monocropping
   c) IT and banking sectors
   d) Mining and industrialization

6. **Which ICAR-supported model is most suitable for commercial farming in water-scarce areas?**
   a) Hydroponics and drip irrigation-based farming
   b) Rice-wheat cropping system
   c) Rainfed jute farming
   d) Wetland rice cultivation
7. **What is the main advantage of ICAR's Integrated Farming System (IFS) models?**
   a) Dependence on single-crop farming
   b) Risk diversification and higher income
   c) Use of only chemical fertilizers
   d) Reduced farm productivity
8. **Which ICAR program promotes agro-processing and value addition?**
   a) National Agriculture Market (e-NAM)
   b) Agri-Business Incubation Centres (ABI)
   c) Mid-Day Meal Scheme
   d) Pradhan Mantri Gram Sadak Yojana
9. **Which of the following is a focus area of ICAR's commercial poultry farming models?**
   a) Indigenous breed improvement
   b) Banning poultry farming
   c) Importing poultry feed
   d) Decreasing poultry farm productivity
10. **ICAR promotes organic farming under which initiative?**
   a) National Program for Organic Production (NPOP)
   b) National Textile Mission
   c) Smart City Program
   d) Digital India Initiative

True/False Questions

1. ICAR supports commercial farming models to enhance farmer incomes and sustainability. **(True)**

2. ICAR discourages the use of modern technology in commercial farming. **(False)** *(ICAR promotes modern farming techniques.)*
3. The National Dairy Research Institute (NDRI) plays a key role in ICAR's dairy-based livelihood models. **(True)**
4. ICAR does not support integrated farming systems. **(False)** *(ICAR actively promotes integrated models for better productivity.)*
5. Agri-Business Incubation Centres (ABI) under ICAR help entrepreneurs develop value-added farm products. **(True)**
6. ICAR's research is only focused on large-scale commercial farming. **(False)** *(ICAR also supports small and marginal farmers.)*
7. Precision farming techniques like hydroponics are part of ICAR-supported farming models. **(True)**
8. ICAR does not work on fishery-based livelihood models. **(False)** *(ICAR has dedicated research institutions for fisheries.)*
9. Krishi Vigyan Kendras (KVKs) help in technology dissemination for commercial farming. **(True)**
10. ICAR encourages diversification of farming systems to reduce risk and improve profitability. **(True)**

19

Commercial Farming-Based Livelihood Models by State Government

Commercial farming is an increasingly vital component of agricultural development strategies in many states. State governments often implement various livelihood models aimed at promoting commercial farming to enhance agricultural productivity, boost local economies, and improve farmers' livelihoods. Here are some common commercial farming-based livelihood models that state governments may employ:

1. Cooperative Farming

- **Description:** Farmers form cooperatives to pool resources, share equipment, and access larger markets. This model strengthens bargaining power and reduces costs.
- **Government Role:** States may provide technical assistance, financial support, and facilitate access to markets.

2. Contract Farming

- **Description:** Farmers enter into agreements with corporations or buyers for the production of specific crops. This model often guarantees a market for the farmers' produce.
- **Government Role:** Facilitation of contracts, monitoring compliance, and providing legal support.

3. Agribusiness Incubators

- **Description:** Establishing incubators to support startups in the agricultural sector, providing mentorship, training, and resources for innovative farming techniques.
- **Government Role:** Funding, infrastructure support, and partnerships with educational institutions.

4. Diversified Farming Systems

- **Description:** Encouraging farmers to engage in multiple agricultural activities (e.g., crop and livestock) to increase income and reduce risk.

- **Government Role:** Providing training on multi-cropping techniques and market access for diverse products.

5. Agri-Value Chain Development

- **Description:** Focusing on different stages of the agricultural value chain, from production to processing and marketing, enhancing overall productivity and profitability.
- **Government Role:** Investment in infrastructure like cold storage, processing facilities, and market linkages.

6. Integrated Farming Systems

- **Description:** Promoting the integration of different agricultural practices, such as crop cultivation, aquaculture, and livestock rearing to optimize resource use and increase income.
- **Government Role:** Providing training and access to inputs and markets.

7. Technology-Driven Farming

- **Description:** Promoting the use of modern technologies, including precision agriculture, drones, and biotech, to enhance productivity.
- **Government Role:** Support for research and development, subsidies for technology adoption, and training for farmers.

8. Sustainable and Organic Farming Initiatives

- **Description:** Programs encouraging organic farming and sustainable agricultural practices that can be marketed at a premium.
- **Government Role:** Certification support, financial incentives, and training on organic practices.

9. Financial Support Schemes

- **Description:** Providing credit facilities, grants, or subsidies to support farmers in transitioning to commercial farming.
- **Government Role:** Partnerships with financial institutions and designing farmer-friendly loan schemes.

10. Market Access and Infrastructure Development

- **Description:** Enhancing the infrastructure for transporting goods, establishing markets, and connecting farmers to consumers directly.
- **Government Role:** Investment in rural roads, marketplaces, and logistics.

Conclusion

State governments can implement a combination of these models tailored to regional agricultural conditions, preferences, and market dynamics. Collaboration with various stakeholders, including private sectors, NGOs, and research institutions, is crucial for the success of these commercial farming initiatives.

QUESTION BANK

Short Answer Questions

1. What are commercial farming-based livelihood models supported by state governments?
2. Name two state government schemes that promote commercial farming.
3. How do state governments support agripreneurs in commercial farming?
4. What role do Farmer Producer Organizations (FPOs) play in state-supported commercial farming models?
5. Name a state that has successfully implemented contract farming for commercial agriculture.
6. How do state governments assist in marketing and value chain development for commercial farming?
7. What is the significance of state-sponsored cold storage and warehousing facilities?
8. How does the state government promote integrated farming systems for commercial agriculture?
9. Which state in India is known for its successful dairy-based livelihood model?
10. How do state governments support organic farming as a commercial enterprise?

Multiple Choice Questions (MCQs)

1. **Which of the following is a key focus of state government-supported commercial farming models?**

   a) Promoting urban industries

   b) Enhancing agricultural productivity and rural employment

   c) Increasing dependence on imports

   d) Banning agricultural exports

2. **Which state has implemented the "Zero Budget Natural Farming" model for commercial agriculture?**

   a) Punjab b) Himachal Pradesh

   c) Andhra Pradesh d) Tamil Nadu

3. **Which state government launched the 'Rythu Bandhu' scheme to provide direct financial assistance to farmers?**

   a) Maharashtra b) Telangana

   c) Karnataka d) West Bengal

4. **Which of the following is a key component of state-sponsored integrated farming systems?**

   a) Industrial manufacturing

   b) Crop-livestock-fishery combination

   c) Software development

   d) Mining operations

5. **Which state is a leader in commercial dairy farming with its 'Amul' cooperative model?**

   a) Kerala b) Gujarat

   c) Bihar d) Rajasthan

6. **Which state has promoted 'Mission Organic Value Chain Development for the North Eastern Region' (MOVCDNER)?**

   a) Assam b) Uttar Pradesh

   c) Madhya Pradesh d) Bihar

7. **Which of the following is a major commercial horticulture initiative by the Himachal Pradesh government?**

   a) Apple Mission b) Cotton Production Scheme

   c) Jute Processing Program d) Fisheries Expansion Plan

8. **The Tamil Nadu government supports aquaculture under which initiative?**

   a) Blue Revolution b) Green Revolution

   c) Textile Export Mission d) Land Reforms Program

9. Which state government launched 'Bhavantar Bhugtan Yojana' to provide price support to farmers?

a) Rajasthan
b) Madhya Pradesh
c) Odisha
d) Uttar Pradesh

10. Which state is well-known for commercial spice farming, particularly black pepper and cardamom?

a) Punjab
b) Kerala
c) Jharkhand
d) Haryana

True/False Questions

1. State governments provide subsidies and financial aid for commercial farming models. **(True)**
2. Only central government schemes support commercial farming; state governments have no role. **(False)** *(State governments actively promote commercial farming through various schemes.)*
3. The Maharashtra government promotes commercial grape farming for wine production. **(True)**
4. State governments do not encourage contract farming for commercial agriculture. **(False)** *(Many states promote contract farming to benefit farmers.)*
5. Cold storage infrastructure funded by state governments helps reduce post-harvest losses. **(True)**
6. The Punjab government has promoted commercial paddy-wheat monoculture over diversified cropping. **(True)**
7. Gujarat is not involved in commercial dairy farming. **(False)** *(Gujarat is home to Amul, India's leading dairy cooperative.)*
8. The Blue Revolution program in Tamil Nadu supports inland and marine aquaculture. **(True)**
9. The Madhya Pradesh government does not provide any price support to farmers. **(False)** *(The state runs the 'Bhavantar Bhugtan Yojana' for price support.)*
10. Organic farming is discouraged by state governments as a commercial enterprise. **(False)** *(Many states actively support organic farming through various initiatives.)*

20

Commercial Farming-Based Livelihood Models by the Other NGOs Across the Country

Across India, various NGOs have developed and implemented innovative commercial farming-based livelihood models aimed at improving the economic status of rural communities. These models often focus on sustainable agricultural practices, value addition, and market linkages. Here are a few noteworthy examples:

1. BAIF Development Research Foundation

- BAIF has implemented programs focused on dairy farming, agroforestry, and watershed management. They provide training, resource support, and market access to farmers, optimizing their income through diversified agricultural activities.

2. Pradan (Professional Assistance for Development Action):

- Pradan works primarily in tribal and backward regions, focusing on enhancing agricultural productivity and promoting non-farm livelihoods. They encourage the use of traditional knowledge, organic farming, and community-based seed banks, facilitating sustainable farming practices.

3. Sambhaavnaa Institute of Public Policy

- This organization focuses on promoting organic farming and the production of high-value crops. They empower local farmers by training them in organic agriculture techniques, enhancing market access for their produce, and developing cooperatives for collective bargaining.

4. Rajasthan's Grameen Vikas Sangh:

- This NGO promotes the cultivation of medicinal plants and organic vegetables. They have established local markets to help farmers sell their products directly to consumers, improving their income while promoting health and sustainability.

5. Aajeevika Bureau

- This organization focuses on informal sector workers and has developed agricultural projects that provide training in sustainable practices, along with facilitating access to government schemes, loans, and markets, thereby enhancing rural livelihoods.

6. Khet Bhakti (Farmer's Friend)

- This initiative emphasizes creating self-reliant farmer communities by promoting organic farming, crop diversification, and integrated farming systems. They also encourage the establishment of cooperatives that help farmers access markets.

7. Digital Green

- By leveraging technology, Digital Green trains farmers in improved agricultural practices and connects them to market opportunities. They use videos and community-led initiatives to disseminate knowledge about sustainable farming.

8. Kissan Ki Awaaz

- A collaboration of various NGOs, this platform focuses on providing farmers with information on weather, pests, farming techniques, and market prices through mobile technology, thereby enabling informed decision-making and increasing profitability.

9. SEWA (Self-Employed Women's Association):

- SEWA emphasizes empowering women through small-scale agriculture and allied activities. They provide training in sustainable practices, financial literacy, and access to markets, improving the livelihoods of women farmers.

10. Barefoot College

- This organization trains rural youth and women in sustainable agricultural practices, water conservation, and entrepreneurship, enabling them to contribute to local economies while ensuring environmental sustainability.

These models not only focus on increasing agricultural productivity but also integrate elements of community development, gender empowerment, and sustainability. By promoting collective action and providing access to resources, these NGOs help rural communities achieve economic independence and resilience.

QUESTION BANK

Short Answer Questions

1. What is the role of NGOs in promoting commercial farming-based livelihood models?
2. Name two NGOs that support commercial farming in India.
3. How do NGOs help in the development of Farmer Producer Organizations (FPOs)?
4. Which NGO is known for promoting organic farming and sustainable agriculture?
5. How do NGOs contribute to value chain development in commercial agriculture?
6. Name an NGO that works in promoting dairy-based livelihoods in rural India.
7. How do NGOs assist farmers in accessing credit and market linkages?
8. What is the significance of NGO-led agro-processing units for farmers?
9. How do NGOs support smallholder farmers in adopting modern technology?
10. Name an NGO that supports tribal communities through commercial farming initiatives.

Multiple Choice Questions (MCQs)

1. Which of the following NGOs is known for promoting organic farming in India?

a) Greenpeace India

b) SEWA (Self-Employed Women's Association)

c) Akshaya Patra Foundation

d) Teach for India

2. Which NGO has played a major role in dairy-based livelihood promotion in India?

a) Pradan

b) Amul Foundation

c) CRY

d) Oxfam India

3. **Which of the following activities is commonly supported by NGOs in commercial farming?**
   a) Promoting IT outsourcing
   b) Establishing Farmer Producer Organizations (FPOs)
   c) Expanding mining operations
   d) Supporting urban real estate

4. **Which NGO works extensively in watershed management for sustainable agriculture?**
   a) World Wildlife Fund (WWF) India
   b) Foundation for Ecological Security (FES)
   c) Child Rights and You (CRY)
   d) Indian Red Cross Society

5. **Which NGO is actively involved in promoting women's participation in farming-based enterprises?**
   a) HelpAge India
   b) SEWA (Self-Employed Women's Association)
   c) Save the Children
   d) PETA India

6. **What is the main benefit of NGO-supported value chains in commercial farming?**
   a) Increased market access for small farmers
   b) Decreased agricultural productivity
   c) Reduction in farmer incomes
   d) Complete dependence on government subsidies

7. **Which NGO promotes sustainable cotton farming in India?**
   a) WWF India b) Pradan
   c) Smile Foundation d) Akshay Patra

8. **Which NGO provides financial literacy programs for small farmers in India?**
   a) BASIX b) Greenpeace
   c) Teach for India d) Amnesty International

9. **Which of the following is a key feature of NGO-led commercial farming models?**
   a) Encouraging sustainable and organic farming
   b) Promoting deforestation for farmland
   c) Reducing the number of Farmer Producer Organizations (FPOs)
   d) Restricting farmers' market access

10. **Which NGO supports beekeeping as a commercial livelihood for rural farmers?**
   a) Under the Mango Tree Society b) PETA India
   c) Greenpeace India d) Child Rights and You (CRY)

True/False Questions

1. NGOs play a significant role in promoting commercial farming-based livelihood models in India. **(True)**
2. Only the government, not NGOs, can establish Farmer Producer Organizations (FPOs). **(False)** *(NGOs actively support and promote FPOs.)*
3. Pradan is an NGO that focuses on promoting sustainable agriculture among small farmers. **(True)**
4. NGOs do not support commercial dairy farming. **(False)** *(Several NGOs promote dairy-based livelihoods.)*
5. SEWA (Self-Employed Women's Association) helps women farmers engage in commercial farming activities. **(True)**
6. WWF India has no involvement in sustainable agriculture. **(False)** *(WWF India promotes sustainable cotton farming and organic agriculture.)*
7. BASIX is an NGO that provides financial and business development services for farmers. **(True)**
8. Greenpeace India promotes genetically modified crops for commercial farming. **(False)** *(Greenpeace India advocates for organic and sustainable farming.)*
9. NGOs help farmers improve post-harvest processing and value addition. **(True)**
10. The Amul Foundation works with NGOs to support dairy farming as a commercial livelihood. **(True)**

21

Case Studies on Different Livelihood Enterprises Associated with the Farming

Livelihood enterprises associated with farming encompass a diverse range of activities that support agricultural income, enhance food security, and provide sustainable livelihoods for farming communities. Below are several case studies from different regions and types of farming-related enterprises:

1. Organic Vegetable Farming in India

- **Location:** Karnataka, India
- **Enterprise Type:** Organic farming
- **Synopsis:** A group of farmers transitioned from conventional agriculture to organic farming, focusing on high-value crops like tomatoes, bell peppers, and leafy greens. They formed a cooperative to share resources, equipment, and knowledge.
- **Impact:** By adopting organic practices, farmers increased their income by 30-50% due to higher market prices for organic vegetables. The cooperative also created a direct supply chain to urban markets, reducing dependence on middlemen.

2. Beekeeping as a Supplementary Enterprise in East Africa

- **Location:** Kenya
- **Enterprise Type:** Beekeeping
- **Synopsis:** Smallholder farmers in rural Kenya began integrating beekeeping into their farming practices. Training sessions were held on sustainable beekeeping techniques and hive management, and the honey produced was marketed locally and abroad.
- **Impact:** Beekeeping not only provided an additional source of income from honey and wax but also improved crop yields due to increased pollination. This diversification helped farmers remain resilient against market fluctuations in their primary crops.

3. Aquaponics in the United States

- **Location:** California, USA
- **Enterprise Type:** Aquaponics
- **Synopsis:** A family farm in California implemented an aquaponics system, integrating fish farming (tilapia) with vegetable cultivation (lettuce and herbs). The system uses the waste produced by fish to fertilize the plants, creating a symbiotic relationship.
- **Impact:** This enterprise improved water efficiency and reduced the need for chemical fertilizers. The farm experienced increased productivity and profitability, with fresh produce sold to local restaurants and farmers' markets.

4. Dairy Cooperative in Bangladesh

- **Location:** Rural Bangladesh
- **Enterprise Type:** Dairy farming
- **Synopsis:** Local farmers formed a dairy cooperative to improve milk production and marketing. They received training on animal husbandry, health management, and quality control and established a milk collection and processing facility.
- **Impact:** The cooperative enhanced the quality and quantity of milk production, providing a steady and better price for members. This led to improved livelihoods, food security, and nutrient availability for local communities.

5. Agroforestry in Brazil

- **Location:** Amazon Rainforest, Brazil
- **Enterprise Type:** Agroforestry
- **Synopsis:** Local farmers adopted agroforestry practices, combining traditional crops (like manioc) with fruit trees and timber species. This approach aimed to restore degraded land while diversifying income sources.
- **Impact:** Apart from ecological benefits such as enhanced biodiversity and carbon sequestration, farmers reported significantly increased household incomes from selling fruits, nuts, and timber products in local and regional markets.

6. Craft Brewing from Barley in Canada

- **Location:** Alberta, Canada
- **Enterprise Type:** Craft brewing

- **Synopsis:** Barley farmers began collaborating with local craft breweries to supply high-quality barley, integrating brewing with traditional farming. The local brewery provided feedback on preferred barley varieties and product quality.
- **Impact:** Farmers gained additional income and created a local market for their produce, while breweries capitalized on local sourcing to enhance their brand and community support.

7. Smallholder Coffee Farms in Colombia

- **Location:** Antioquia, Colombia
- **Enterprise Type:** Specialty coffee production
- **Synopsis:** Smallholder farmers organized into associations to improve coffee production quality and adopt sustainable practices. They received training on processing techniques to meet the growing demand for specialty coffee.
- **Impact:** The farmers successfully accessed international markets and gained certifications (e.g., Fair Trade, Organic), leading to improved prices, better living conditions, and increased community investment.

Conclusion

These case studies highlight the importance of diversification, cooperatives, sustainable practices, and market access in improving the livelihoods of farming communities. By adopting innovative approaches, farmers can enhance their income, support food security, and build resilience against economic and environmental challenges.

QUESTION BANK

Short Answer Questions

1. What is a farming-based livelihood enterprise?
2. Give an example of a successful Farmer Producer Organization (FPO) in India.
3. How does dairy farming contribute to rural livelihoods?
4. Name a case study of an agro-processing unit that benefited small farmers.
5. How has organic farming improved farmer incomes in India?
6. Mention one successful beekeeping enterprise in India.
7. How do integrated farming systems enhance farmer incomes?
8. What role do women-led enterprises play in agricultural livelihoods?

9. Give an example of a cooperative society that supports farming-based enterprises.
10. How has aquaculture helped rural communities in coastal areas?

Multiple Choice Questions (MCQs)

1. Which of the following is a successful dairy cooperative in India?

a) Amul b) NAFED

c) ITC d) Reliance Fresh

2. Which farming-based enterprise is commonly practiced in Sundarbans for livelihood?

a) Poultry farming b) Beekeeping

c) Shrimp aquaculture d) Cotton cultivation

3. Which Indian state is famous for promoting organic farming through Sikkim Organic Mission?

a) Punjab b) Gujarat

c) Sikkim d) Tamil Nadu

4. Which organization helps small farmers establish and scale up Farmer Producer Organizations (FPOs)?

a) NABARD b) WHO

c) WTO d) RBI

5. Which of the following is an example of an integrated farming system?

a) Poultry cum fish farming b) IT consultancy

c) Automobile manufacturing d) Textile production

6. Which state has a successful case study on silk farming as a livelihood enterprise?

a) Bihar b) Karnataka

c) Rajasthan d) Kerala

7. Which of the following enterprises is related to post-harvest value addition in farming?

a) Cold storage facility b) Mobile app development

c) Cement production d) Road construction

8. **Which NGO has promoted commercial beekeeping for rural farmers in India?**
   a) Greenpeace
   b) Under The Mango Tree Society
   c) Oxfam
   d) World Wildlife Fund (WWF)

9. **Which crop has been successfully linked to commercial farming through contract farming in Punjab?**
   a) Wheat
   b) Basmati rice
   c) Cotton
   d) Sugarcane

10. **Which case study highlights the success of horticultural farming as a livelihood enterprise?**
    a) Apple orchards in Himachal Pradesh
    b) Tea plantations in Rajasthan
    c) Coffee cultivation in Punjab
    d) Rubber plantations in Gujarat

True/False Questions

1. Amul is an example of a successful farming-based livelihood enterprise. **(True)**
2. Integrated farming systems involve multiple farm-based enterprises working together. **(True)**
3. Organic farming has not contributed to improving rural incomes in India. **(False)** *(Organic farming has helped increase farmer incomes.)*
4. Shrimp aquaculture has provided sustainable livelihoods in the Sundarbans. **(True)**
5. Poultry farming is an unsuccessful model for rural livelihood development. **(False)** *(Poultry farming is a major source of income for rural households.)*
6. Farmer Producer Organizations (FPOs) help small farmers access better markets. **(True)**
7. Silk farming is not a viable commercial enterprise in India. **(False)** *(Silk farming is a major livelihood source in Karnataka and other states.)*
8. Cold storage infrastructure does not play a role in post-harvest management. **(False)** *(Cold storage helps reduce post-harvest losses.)*

9. Contract farming has helped improve commercial farming in some regions of India. **(True)**
10. Apple farming in Himachal Pradesh is a successful case study in horticulture-based livelihoods. **(True)**

22

Risk Factors in Farming-Based Livelihood Systems

Farming-based livelihood systems are complex and variable, influenced by a wide range of factors that can pose risks to agricultural production and the well-being of farming communities. Here are some key risk factors:

Environmental Risks

1. **Climate Change**: Increased temperatures, altered precipitation patterns, and extreme weather events (droughts, floods, hurricanes) can dramatically affect crop yields and livestock health.
2. **Soil Degradation**: Erosion, loss of soil fertility, and contamination can reduce agricultural productivity over time.
3. **Water Availability**: Both the scarcity and pollution of water sources can impact irrigation and livestock, leading to reduced yields and livestock health.
4. **Pest and Disease Outbreaks**: Invasive insects, diseases in crops or livestock, and a lack of pest management practices can threaten production.

Economic Risks

1. **Market Fluctuations**: Volatility in prices for crops and livestock can affect farmers' income stability. Dependence on a single crop can exacerbate risks.
2. **Access to Credit**: Limited access to financial services can hinder the ability to invest in necessary inputs or recover from crop failures.
3. **Rising Input Costs**: Increases in costs for seeds, fertilizers, and agricultural machinery can affect profit margins.
4. **Trade Barriers**: Changes in policy, tariffs, and trade agreements can affect market access and profitability.

Social Risks

1. **Land Tenure Issues**: Unclear land rights and insecurity can discourage investment in sustainable practices and improvements.
2. **Labor Availability**: Fluctuations in labor availability, particularly seasonal labor shortages, can impact planting and harvesting.
3. **Community Conflicts**: Land disputes, competition over resources, and social tensions can undermine community cohesion and affect productivity.

Technological Risks

1. **Access to Technology**: Limited access to modern farming technology and practices can hinder productivity improvements and adaptation to changing conditions.
2. **Biodiversity Loss**: Over-reliance on monoculture and traditional practices can lead to a loss of genetic diversity, making crops and livestock more vulnerable to pests and diseases.

Policy and Governance Risks

1. **Regulatory Changes**: Sudden changes in agricultural policies, subsidies, or environmental regulations can create uncertainty for farmers.
2. **Lack of Support Services**: Insufficient agricultural extension services, training, and research can prevent farmers from accessing information needed to improve practices.

Health Risks

1. **Public Health Issues**: Diseases, both in humans and livestock, can disrupt farming activities and livelihoods.
2. **Food Safety Concerns**: Contamination and safety issues can affect market access for agricultural products.

Mitigation Strategies

To address these risks, farmers and stakeholders can adopt various strategies, such as:

- Diversifying crops and income sources
- Implementing sustainable agricultural practices
- Building strong community networks
- Investing in technology and infrastructure
- Enhancing access to markets and financial services
- Engaging in policy advocacy for supportive agricultural policies.

Understanding and addressing these risk factors is essential for the sustainability and resilience of farming-based livelihood systems.

QUESTION BANK

Short Answer Questions

1. What are the major risk factors affecting farming-based livelihoods?
2. How does climate change impact agricultural productivity?
3. What role do pests and diseases play in farming risks?
4. How does market price fluctuation affect farmers' income?
5. Mention one financial risk faced by small-scale farmers.
6. How does soil degradation affect long-term farm productivity?
7. What are the socio-economic risks associated with farming?
8. How can farmers mitigate risks related to extreme weather events?
9. What is the role of insurance in managing farming risks?
10. How does lack of access to credit increase farming risks?

Multiple Choice Questions (MCQs)

1. Which of the following is a major climatic risk in farming?

a) Political instability　b) Droughts and floods

c) Road construction　d) Social media trends

2. What is the most common pest-related risk in agriculture?

a) High land prices　b) Insect infestations

c) Lack of farm labor　d) Poor internet connectivity

3. Which of the following is an economic risk for farmers?

a) Stable market prices　b) High input costs

c) Good soil health　d) Effective irrigation systems

4. How can farmers reduce market-related risks?

a) By depending on a single buyer

b) By engaging in contract farming

c) By ignoring market trends

d) By selling only during peak seasons

5. Which of these is an environmental risk for farmers?

a) Increasing access to technology　b) Soil erosion

c) Improved irrigation facilities　d) Higher farm subsidies

6. **Which factor increases the financial risks for farmers?**
   a) Government subsidies
   b) Easy access to credit
   c) High interest rates on loans
   d) Crop diversification
7. **Which of the following farming practices can help reduce risk?**
   a) Monocropping
   b) Diversification of crops
   c) Ignoring soil fertility
   d) Overuse of chemical fertilizers
8. **What is the purpose of crop insurance in farming?**
   a) To increase pesticide usage
   b) To compensate farmers for crop losses
   c) To encourage deforestation
   d) To eliminate farm labor costs
9. **Which of the following is an example of a market risk?**
   a) Unexpected drought
   b) Declining demand for a particular crop
   c) Proper post-harvest storage
   d) Timely access to fertilizers
10. **Which policy can help farmers manage risks in farming?**
   a) MSP (Minimum Support Price)
   b) Increasing middlemen in the supply chain
   c) Restricting credit facilities
   d) Eliminating modern technology

True/False Questions

1. Climate change is not a significant risk factor for farmers. **(False)**
2. Diversified cropping systems help reduce farming risks. **(True)**
3. Soil degradation has no impact on farming-based livelihoods. **(False)**
4. Market fluctuations can cause serious income instability for farmers. **(True)**
5. High dependency on a single crop reduces financial risks for farmers. **(False)**

6. Government policies such as MSP help stabilize farmers' income. **(True)**
7. Pests and diseases are not major threats to crop production. **(False)**
8. Crop insurance provides financial security to farmers against losses. **(True)**
9. Poor post-harvest storage does not contribute to economic risks in farming. **(False)**
10. Water scarcity is a key environmental risk in farming. **(True)**

23

Success Factors in Farming-Based Livelihood Systems

Farming-based livelihood systems can be influenced by a variety of risk and success factors. Understanding these can help farmers and policymakers develop strategies to enhance sustainability and resilience in agricultural communities. Here are key points outlining both risk and success factors.

Risk Factors

1. Climate Change and Weather Variability:

- Unpredictable weather patterns, droughts, floods, and shifting growing seasons can adversely affect crop yields.

2. Market Fluctuations

- Price volatility for agricultural products can impact income stability. Changes in demand and supply dynamics can also affect profitability.

3. Pest and Disease Outbreaks

- Infestations of pests or diseases can lead to significant crop losses and increased costs for farmers.

4. Access to Resources

- Limited access to land, water, credit, and agricultural inputs (seeds, fertilizers, equipment) can hinder productivity.

5. Soil Degradation

- Poor soil health due to overuse of chemicals, erosion, or unsustainable farming practices can reduce agricultural output.

6. Policy and Institutional Challenges

- Inadequate support from government policies, lack of infrastructure, and weak institutional frameworks can prevent farmers from thriving.

7. Social and Economic Inequities

- Marginalized farmers may lack access to education, technology, and markets, limiting their capacity to improve their livelihoods.

8. Health and Labor Issues

- Illnesses or labor shortages can affect farm operations, leading to production losses.

9. Environmental Degradation

- Over-exploitation of natural resources can lead to long-term sustainability issues, impacting farm viability.

Success Factors

1. Diversification

- Diversifying crops and livestock can reduce risk and enhance resilience to market and climate changes.

2. Access to Technology

- Utilizing modern farming techniques, precision agriculture, and digital tools can improve productivity and efficiency.

3. Education and Training

- Knowledge and skill development in sustainable farming practices can lead to better decision-making and increased yields.

4. Strong Market Links

- Establishing cooperatives or direct market access can enhance farmers' bargaining power and income security.

5. Sustainable Practices

- Implementing conservation agriculture, agroforestry, and organic farming can contribute to long-term environmental health and resilience.

6. Financial and Risk Management Tools

- Access to credit, insurance, and savings programs can help farmers manage financial risks and invest in improvements.

7. Community Support and Networks

- Building social capital through community organizations can provide support, resources, and knowledge-sharing.

8. Government and Policy Support

- Effective agricultural policies, subsidies, and infrastructure development can foster a more conducive environment for farming.

9. Climate-Resilient Strategies

- Adopting practices that enhance soil health and water conservation can help mitigate the impacts of climate change.

10. Research and Innovation

- Ongoing research and the development of new agricultural technologies can improve crop varieties and farming methods.

By actively managing risk factors and leveraging success factors, farming-based livelihood systems can become more sustainable and resilient, ensuring food security and better livelihoods for farming communities.

QUESTION BANK

Short Answer Questions

1. What are the key success factors in farming-based livelihood systems?
2. How does access to credit contribute to the success of farming enterprises?
3. Why is crop diversification important for sustainable farming livelihoods?
4. How does the adoption of modern technology improve farm productivity?
5. What role do Farmer Producer Organizations (FPOs) play in improving farmer income?
6. Why is access to markets essential for the success of farming livelihoods?
7. How does water management impact the success of a farming system?
8. Name one government scheme that supports farming-based livelihoods in India.
9. What is the significance of value addition in agricultural products?
10. How does climate resilience contribute to long-term farming success?

Multiple Choice Questions (MCQs)

1. Which of the following is a major factor in the success of farming livelihoods?

a) Lack of infrastructure

b) Adoption of modern agricultural practices

c) High dependence on a single crop

d) Excessive chemical fertilizer use

2. **Which of the following helps farmers get better prices for their produce?**

   a) Middlemen exploitation
   b) Direct market linkages
   c) Selling at low prices
   d) Increasing input costs

3. **What is one of the most effective ways to reduce farming risks and improve success?**

   a) Growing only one type of crop
   b) Practicing integrated farming systems
   c) Avoiding access to credit
   d) Using outdated farming techniques

4. **Which of these is an important financial factor for successful farming?**

   a) Access to credit and subsidies
   b) Lack of investment in farming
   c) High interest rates on loans
   d) Selling crops without market knowledge

5. **Which of the following is a technological success factor in farming?**

   a) Use of improved seeds and precision farming
   b) Avoiding modern irrigation methods
   c) Overuse of pesticides
   d) Dependency on outdated tools

6. **What is the benefit of forming Farmer Producer Organizations (FPOs)?**

   a) Reducing farmer cooperation
   b) Helping farmers get better market prices
   c) Increasing individual investment risk
   d) Limiting government support

7. **Why is soil health management important for farming success?**

   a) It reduces crop productivity
   b) It improves yield and sustainability
   c) It increases dependency on synthetic fertilizers
   d) It makes farming more expensive

8. Which farming method is best suited for long-term success?

a) Monocropping

b) Sustainable and organic farming

c) Excessive pesticide use

d) Overgrazing of land

9. Which government scheme in India supports farm mechanization?

a) PM-KISAN

b) Kisan Credit Card (KCC)

c) Sub-Mission on Agricultural Mechanization (SMAM)

d) Mid-Day Meal Scheme

10. Which factor contributes to better resilience in farming-based livelihoods?

a) Dependence on a single crop

b) Climate-smart agricultural practices

c) Ignoring market trends

d) Not using modern technology

True/False Questions

1. Access to finance and credit is not important for farming success. **(False)**
2. Crop diversification helps reduce risks and improves farmer income. **(True)**
3. Proper irrigation and water management are key success factors in farming. **(True)**
4. Use of modern technology, such as precision farming, does not affect farming success. **(False)**
5. Value addition in agriculture (e.g., food processing) increases farmer income. **(True)**
6. Lack of market linkages improves the profitability of farming enterprises. **(False)**
7. Government schemes and policies play a role in enhancing farming livelihoods. **(True)**
8. Monocropping is a sustainable long-term farming practice. **(False)**
9. Integrated farming systems help improve soil health and farm productivity. **(True)**
10. Climate resilience strategies are essential for the sustainability of farming-based livelihoods. **(True)**

24

Schemes and Programs by Central Government and Public Involved in Promotion of Farming-Based Livelihood Opportunities

The Government of India has launched several schemes and programs aimed at promoting farming-based livelihood opportunities, enhancing agricultural productivity, and supporting rural development. Here are some of the key schemes and initiatives:

1. Pradhan Mantri Kisan Samman Nidhi (PM-KISAN)

- **Objective**: Direct income support to small and marginal farmers.
- **Features**: Provides □6,000 annually, payable in three installments, to help farmers purchase inputs and sustain their livelihoods.

2. Pradhan Mantri Fasal Bima Yojana (PMFBY)

- **Objective**: To provide risk coverage to farmers for crop loss.
- **Features**: Insurance for crops against natural calamities, pests, and diseases to ensure financial stability.

3. Soil Health Card Scheme

- **Objective**: To promote balanced use of fertilizers.
- **Features**: Issuance of soil health cards to farmers every 2 years, providing information on nutrient status and recommendations for fertilizers.

4. National Agricultural Market (eNAM)

- **Objective**: To facilitate online trading of agricultural products.
- **Features**: A unified platform for farmers, traders, and buyers to sell and purchase agricultural produce via online auctions.

5. Pradhan Mantri Krishi Sinchai Yojana (PMKSY)

- **Objective**: To expand irrigated land and ensure access to water for agriculture.

- **Features**: Promotion of micro-irrigation (drip and sprinkler systems) and watershed development.

6. Rural Infrastructure Development Fund (RIDF)

- **Objective**: To improve rural infrastructure.
- **Features**: Finance for projects that enhance farm productivity and rural livelihoods.

7. Digital India Initiative

- **Objective**: To promote digital literacy among farmers.
- **Features**: Use of mobile technologies and the internet to provide farmers with information on weather, markets, and modern farming techniques.

8. National Livestock Mission

- **Objective**: To promote sustainable livestock development.
- **Features**: Enhancements in livestock productivity and support to farmers for rearing livestock.

9. Self-Employment and Skill Development Programs

- **Objective**: To enhance farmers' skills and promote alternative livelihoods.
- **Features**: Initiatives like the Deen Dayal Upadhyaya Grameen Kaushalya Yojana focus on skill development in agricultural practices and related fields.

10. PM Awas Yojana - Gramin

- **Objective**: Provide housing to rural poor, enhancing living conditions.
- **Features**: Aims to provide a pucca house with basic amenities, which can also support agricultural livelihoods.

11. Start-up India Scheme

- **Objective**: To promote entrepreneurship in agriculture.
- **Features**: Support for agritech startups providing innovative solutions to farmers.

12. Atmanirbhar Bharat Abhiyan (Self-Reliant India Mission)

- **Objective**: To boost local production and self-sufficiency, especially in agriculture.
- **Features**: Emphasizes integrating local economies and promoting local products.

13. National Organic Farming Policy

- **Objective**: To promote organic farming practices.
- **Features**: Encourages farmers to adopt organic farming through training and market support.

Conclusion

These schemes aim to empower farmers and improve their livelihoods by providing financial support, insurance, skill development, modern practices, and market access. Ensuring the sustainability and growth of agriculture remains a primary focus for the government, with continuous efforts to enhance policies and programs in this sector.

QUESTION BANK

Short Answer Questions

1. Name any two central government schemes that promote farming-based livelihoods in India.
2. What is the objective of the Pradhan Mantri Kisan Samman Nidhi (PM-KISAN) scheme?
3. How does the National Rural Livelihoods Mission (NRLM) support farmers?
4. What is the main focus of the Rashtriya Krishi Vikas Yojana (RKVY)?
5. How does the Kisan Credit Card (KCC) scheme benefit farmers?
6. What is the role of NABARD in supporting farming-based livelihood programs?
7. What kind of support is provided under the Sub-Mission on Agricultural Mechanization (SMAM)?
8. How does the PM Fasal Bima Yojana (PMFBY) help farmers?
9. What is the key focus of the Paramparagat Krishi Vikas Yojana (PKVY)?
10. Which central government scheme focuses on organic farming and sustainable agriculture?

Multiple Choice Questions (MCQs)

1. Which scheme provides direct income support to farmers in India?

a) MGNREGA b) PM-KISAN

c) National Food Security Mission d) Soil Health Card Scheme

2. **What is the main objective of PM Fasal Bima Yojana (PMFBY)?**
   a) Provide free seeds to farmers
   b) Offer crop insurance against natural calamities
   c) Encourage organic farming
   d) Develop rural markets

3. **Which of the following schemes supports women farmers and self-help groups?**
   a) NRLM (National Rural Livelihoods Mission)
   b) Kisan Credit Card
   c) PM-KISAN
   d) Rashtriya Krishi Vikas Yojana

4. **What does the Soil Health Card Scheme aim to improve?**
   a) Irrigation facilities
   b) Farmers' bank accounts
   c) Soil fertility and nutrient management
   d) Crop insurance coverage

5. **Which scheme promotes farm mechanization in India?**
   a) Rashtriya Krishi Vikas Yojana
   b) Sub-Mission on Agricultural Mechanization (SMAM)
   c) PM-KISAN
   d) National Horticulture Mission

6. **Which financial institution plays a key role in funding rural and agricultural development projects?**
   a) SEBI b) NABARD
   c) RBI d) NITI Aayog

7. **The Paramparagat Krishi Vikas Yojana (PKVY) focuses on which type of farming?**
   a) Monocropping b) Organic farming
   c) Conventional farming d) Aquaculture

8. Which of the following is a key feature of the e-NAM platform?

a) Providing crop insurance

b) Facilitating online trading of agricultural products

c) Offering direct cash benefits to farmers

d) Distributing free seeds

9. Which government program is aimed at improving irrigation infrastructure?

a) PM Fasal Bima Yojana

b) Pradhan Mantri Krishi Sinchayee Yojana (PMKSY)

c) Kisan Credit Card

d) NRLM

10. What is the main benefit of the Kisan Credit Card (KCC) scheme?

a) Free land for farmers

b) Easy access to credit at lower interest rates

c) Free crop insurance

d) Direct government employment

True/False Questions

1. PM-KISAN provides financial support to farmers through direct bank transfers. **(True)**
2. The National Rural Livelihoods Mission (NRLM) is only for urban entrepreneurs. **(False)**
3. Soil Health Card Scheme helps farmers by improving soil fertility. **(True)**
4. PM Fasal Bima Yojana does not cover losses due to natural calamities. **(False)**
5. NABARD primarily focuses on urban business development. **(False)**
6. Paramparagat Krishi Vikas Yojana (PKVY) promotes chemical-based farming. **(False)**
7. e-NAM is an online platform for selling agricultural produce. **(True)**
8. The Kisan Credit Card (KCC) scheme allows farmers to get short-term credit. **(True)**
9. Pradhan Mantri Krishi Sinchayee Yojana (PMKSY) focuses on improving irrigation. **(True)**
10. Rashtriya Krishi Vikas Yojana (RKVY) supports holistic agricultural development. **(True)**

25

Schemes and Programs by State Government Organizations Involved in Promotion of Farming-Based Livelihood Opportunities

State governments and public organizations in various regions implement a multitude of schemes and programs to promote farming-based livelihood opportunities. These initiatives often aim to enhance agricultural productivity, improve rural livelihoods, provide financial assistance, and support sustainable farming practices. Here are some common types of schemes and examples that can be found in different Indian states:

1. Financial Assistance Programs

- **Kisan Credit Card (KCC) Scheme**: Provides farmers with timely and adequate credit for agricultural activities, helping them meet their cash flow needs during the farming season.
- **Interest Subsidy Schemes**: Certain states offer schemes that provide interest subsidies on agricultural loans to encourage borrowing for farming activities.

2. Crop Insurance Schemes

- **Pradhan Mantri Fasal Bima Yojana (PMFBY)**: A central scheme adopted by many states that provides insurance coverage and financial support to farmers in the event of crop failure due to natural calamities.

3. Skill Development and Training Programs

- **Integrated Skill Development Scheme (ISDS)**: Aimed at enhancing the skills of farmers and rural youth in various agricultural practices, including modern farming techniques, post-harvest management, and value-added processing.
- **State Agricultural Universities (SAUs)**: Often conduct training programs for farmers on the latest techniques in crop production, pest management, and organic farming.

4. **Support for Organic Farming**

- **Organic Farming Policy**: Various states have developed policies that promote organic farming through subsidies, certification processes, and training programs to encourage sustainable agricultural practices.

5. **Market Access Initiatives**

- **State Agricultural Marketing Boards**: These organizations typically implement schemes to facilitate better access to markets for farmers, including setting up "mandis" (markets) and providing information on market prices.
- **Direct benefit transfer schemes**: Some states have started programs to link farmers directly with consumers, bypassing intermediaries to ensure they receive fair prices for their produce.

6. **Infrastructure Development Programs**

- **Rural Infrastructure Development Fund (RIDF)**: Used by many states to build necessary infrastructure like irrigation facilities, roads, and storage facilities to support agricultural activities.
- **Warehouse Subsidy Schemes**: To provide financial assistance for the construction of godowns and cold storage facilities to help farmers store and preserve their produce.

7. **Women Empowerment in Agriculture**

- **Self-Help Groups (SHGs)**: Programs supporting women in forming SHGs to promote entrepreneurship in agriculture-related businesses such as dairy, poultry, or organic farming.
- **Women in Agriculture Programs**: Various states run specific programs aimed at empowering women in agriculture through training, access to credit, and capacity-building initiatives.

8. **Agri-Entrepreneurship Development Programs**

- **Start-up India Scheme**: Agripreneurs can avail benefits under this scheme for innovation in agriculture, including access to funding and mentoring.
- **National Agro-Industry Policy**: Encourages agri-business initiatives by providing various support mechanisms for agri-entrepreneurs.

9. **Sustainable and Climate-Resilient Agriculture Initiatives**

- **Watershed Development Programs**: Initiatives focused on sustainable land management and water conservation techniques to improve agricultural productivity while preserving the environment.

- **National Adaptation Fund for Climate Change (NAFCC)**: Support for projects that aim to build resilience among farmers against climate change impacts.

Conclusion

The specific schemes and programs may vary from state to state, reflecting local agricultural challenges and opportunities. Hence, farmers are encouraged to connect with local government agricultural departments, cooperative societies, and NGOs for the latest information on available support mechanisms in their regions.

QUESTION BANK

Short Answer Questions

1. Name any two state government schemes that support farming-based livelihoods.
2. How do state agricultural universities contribute to promoting sustainable farming?
3. What is the primary objective of state-specific organic farming programs?
4. Which state government scheme focuses on improving irrigation facilities for farmers?
5. How do cooperative societies in different states help farmers in marketing their produce?
6. What kind of financial support is provided by state governments to small and marginal farmers?
7. How do state rural livelihood missions contribute to farming-based livelihoods?
8. Which state has launched a specific program for promoting horticulture-based livelihoods?
9. How do state-level subsidies for farm mechanization help improve agricultural productivity?
10. Name a state government initiative that supports dairy farmers in rural areas.

Multiple Choice Questions (MCQs)

1. Which Indian state launched the 'Rythu Bandhu' scheme to provide direct financial support to farmers?

a) Maharashtra b) Telangana

c) West Bengal d) Uttar Pradesh

2. **The 'Krushak Assistance for Livelihood and Income Augmentation' (KALIA) scheme is implemented by which state?**
   a) Bihar b) Odisha
   c) Punjab d) Kerala

3. **What is the main focus of the 'Mukhyamantri Krishi Ashirwad Yojana' in Jharkhand?**
   a) Providing free fertilizers
   b) Direct income support to farmers
   c) Free irrigation facilities
   d) Promotion of organic farming

4. **Which state government introduced the 'Bhavantar Bhugtan Yojana' to protect farmers from price fluctuations?**
   a) Madhya Pradesh b) Rajasthan
   c) Gujarat d) Tamil Nadu

5. **The 'Krishi Bhagya' scheme in Karnataka mainly focuses on:**
   a) Soil conservation
   b) Rainwater harvesting for dryland farmers
   c) Providing free hybrid seeds
   d) Promoting chemical fertilizers

6. **Which of the following is a state-run cooperative society that helps farmers with fair pricing and procurement?**
   a) FCI (Food Corporation of India) b) NAFED
   c) AMUL d) ICAR

7. **Which state launched the 'Atal Krishi Jyoti Yojana' to provide electricity for agricultural use?**
   a) Haryana b) Gujarat
   c) Rajasthan d) Punjab

8. **The 'Rajiv Gandhi Kisan Nyay Yojana' was introduced by which state to provide input subsidies to farmers?**
   a) Chhattisgarh b) Assam
   c) Karnataka d) Bihar

9. Which state has implemented the 'Magel Tyala Shettale' scheme to support water conservation for farmers?

a) Maharashtra
b) Himachal Pradesh
c) Uttar Pradesh
d) West Bengal

10. The 'Uzhavan App' was launched by which state government to provide real-time farming-related information to farmers?

a) Andhra Pradesh
b) Tamil Nadu
c) Kerala
d) Telangana

True/False Questions

1. The 'Rythu Bandhu' scheme provides financial assistance to farmers before sowing crops. **(True)**
2. Odisha's KALIA scheme only benefits large landholding farmers. **(False)**
3. The Bhavantar Bhugtan Yojana helps protect farmers from market price fluctuations. **(True)**
4. Karnataka's 'Krishi Bhagya' scheme focuses on distributing chemical fertilizers. **(False)**
5. AMUL is an example of a successful state-supported dairy cooperative society. **(True)**
6. The 'Mukhyamantri Krishi Ashirwad Yojana' provides direct income support to farmers in Jharkhand. **(True)**
7. The 'Rajiv Gandhi Kisan Nyay Yojana' provides financial support only to sugarcane farmers. **(False)**
8. The 'Atal Krishi Jyoti Yojana' in Gujarat provides subsidized solar-powered irrigation pumps to farmers. **(True)**
9. The 'Magel Tyala Shettale' scheme in Maharashtra focuses on rainwater harvesting for farm ponds. **(True)**
10. The 'Uzhavan App' is a mobile application launched by the Kerala government for farmers. **(False)**

26

Schemes and Programs by State Government Organizations Involved in Promotion of Farming-Based Livelihood Opportunities

State governments in India implement various schemes and programs aimed at promoting farming-based livelihood opportunities. These initiatives often target different aspects of agriculture, including production, marketing, technology, and skill development. While the specific schemes can vary by state, here are some common types of programs and initiatives that state governments might implement:

1. Crop Support Schemes

- **Minimum Support Price (MSP):** Several state governments announce MSP for various crops to ensure farmers receive a fair price for their produce.
- **Price Stabilization Funds:** These are meant to support farmers during periods of low prices and ensure profitability.

2. Skill Development Programs

- **Agricultural Skill Development Centers:** Training programs aimed at enhancing the skills of farmers in modern agricultural practices, sustainable farming, and post-harvest management.
- **Workshops and Seminars:** Regular sessions are conducted for farmers to update them on new technologies and practices in agriculture.

3. Subsidies and Financial Assistance

- **Seed and Fertilizer Subsidies:** Financial help for farmers to purchase high-quality seeds and fertilizers.
- **Farm Machinery Subsidies:** Support for purchasing tractors and other essential farm equipment.

4. Irrigation and Water Management Programs

- **Micro Irrigation Schemes:** Promoting drip and sprinkler irrigation to conserve water and improve crop yield.
- **Rainwater Harvesting Initiatives:** Projects to enhance water availability for agricultural use.

5. Access to Credit

- **Kisan Credit Cards (KCC):** Facilitation of loans at subsidized interest rates for farmers.
- **Cooperative Banks and Societies:** Encouraging farmers to utilize local financial institutions for better access to credit.

6. Market Development Initiatives

- **Agricultural Markets**: Establishment of mandis (markets) where farmers can sell their produce directly.
- **Direct Beneficiary Transfer Schemes:** Financial assistance to farmers through direct bank transfers to eliminate middlemen.

7. Organic Farming Promotion

- **Certification and Training Programs:** Initiatives to support organic farming practices and organic certification for farmers.
- **Support for Organic Inputs:** Subsidies or assistance for organic fertilizers and pesticides.

8. Dairy, Poultry, and Livestock Programs

- **Integrated Livestock Development Programs:** Initiatives supporting dairy, poultry, and other livestock farming.
- **Animal Health Care Schemes:** Access to veterinary services and vaccination for livestock.

9. Agri-Entrepreneurship Development

- **Start-Up and Innovation Funds:** Support for young entrepreneurs in agritech and related sectors.
- **Incubation Centers:** Establishment of centers to nurture agri-startups and provide necessary mentorship.

10. Climate-Resilient Farming Initiatives

- **Agroforestry Programs:** Encouraging practices that combine agriculture with forestry to enhance sustainability.

- **Research and Development Programs:** Initiatives to develop drought-resistant and pest-resistant crop varieties.

11. Women Empowerment Schemes in Agriculture

- **Women Farmer's Cooperative Societies:** Encouraging women to engage in farming and earn livelihoods through cooperatives.
- **Skill Training for Women:** Special training programs aimed at empowering women in agriculture.

12. Digital Initiatives

- **Mobile Apps for Farmers:** Apps providing information on weather, market prices, and farming techniques.
- **E-NAM (National Agriculture Market):** Online trading platform for agricultural commodities to ensure better market access.

Examples of State-Specific Schemes

- **Maharashtra:** The 'Jal Yukta Shivar Abhiyan' aims at improving rainwater harvesting.
- **Punjab:** The 'M-Kisan' initiative promotes mobile connectivity for information sharing.
- **Tamil Nadu:** The 'Nammai Kaanalum' program focuses on sustainable farming practices.

Conclusion

These schemes and programs are designed to enhance the livelihoods of farmers, improve agricultural productivity, and create a more sustainable farming environment. It is advisable for farmers to stay informed about the specific programs available in their respective states and to take advantage of them for better economic outcomes.

QUESTION BANK

Short Answer Questions

1. Name two state government schemes that provide direct financial assistance to farmers.
2. What is the main objective of the Odisha government's KALIA scheme?
3. Which state launched the "Rythu Bandhu" scheme to support farmers?
4. How does the Bhavantar Bhugtan Yojana in Madhya Pradesh help farmers?
5. What is the focus of the Krishi Bhagya scheme in Karnataka?

6. Name a state-specific scheme that promotes organic farming.
7. How does the Mukhyamantri Krishi Ashirwad Yojana benefit farmers in Jharkhand?
8. Which state government launched the Atal Krishi Jyoti Yojana to provide electricity for farming?
9. What kind of farming-related support does the Uzhavan App provide to farmers?
10. Which state government launched the "Rajiv Gandhi Kisan Nyay Yojana" to provide financial support to farmers?

Multiple Choice Questions (MCQs)

1. Which state launched the "Rythu Bandhu" scheme to provide financial aid to farmers?

a) Maharashtra b) Telangana

c) West Bengal d) Gujarat

2. The KALIA scheme in Odisha primarily aims to support:

a) Large landholding farmers b) Small and marginal farmers

c) Only horticulture farmers d) Corporate farming

3. What is the main objective of the "Bhavantar Bhugtan Yojana" in Madhya Pradesh?

a) Providing free seeds to farmers

b) Offering minimum price support in case of market price drop

c) Distributing farm machinery

d) Supporting organic farming

4. The "Krishi Bhagya" scheme in Karnataka is designed to help farmers with:

a) Organic certification

b) Rainwater harvesting for dryland agriculture

c) Loan waivers

d) Dairy farming

5. Which state introduced the "Mukhyamantri Krishi Ashirwad Yojana" to provide direct financial assistance to farmers?

a) Bihar b) Jharkhand

c) Rajasthan d) Punjab

6. Which of the following state-supported initiatives focuses on dairy farming?

a) AMUL b) FCI

c) NABARD d) PM-KISAN

7. The "Atal Krishi Jyoti Yojana" is a state government scheme that provides electricity for farming in

a) Haryana b) Gujarat

c) Uttar Pradesh d) Assam

8. Which state government introduced the "Rajiv Gandhi Kisan Nyay Yojana" to provide direct cash benefits to farmers?

a) Chhattisgarh b) Karnataka

c) Bihar d) West Bengal

9. Which state has implemented the "Magel Tyala Shettale" scheme to support water conservation for farming?

a) Maharashtra b) Tamil Nadu

c) Himachal Pradesh d) Madhya Pradesh

10. The "Uzhavan App," which provides farming-related information to farmers, was launched by which state government?

a) Kerala b) Tamil Nadu

c) Telangana d) Andhra Pradesh

True/False Questions

1. The "Rythu Bandhu" scheme provides financial assistance to farmers before sowing crops. **(True)**
2. The "KALIA" scheme only benefits large-scale commercial farmers. **(False)**
3. The "Bhavantar Bhugtan Yojana" helps farmers by protecting them from market price fluctuations. **(True)**
4. The "Krishi Bhagya" scheme in Karnataka provides free fertilizers to farmers. **(False)**
5. AMUL is an example of a successful state-supported dairy cooperative society. **(True)**
6. The "Mukhyamantri Krishi Ashirwad Yojana" provides direct financial support to farmers in Jharkhand. **(True)**

7. The “Rajiv Gandhi Kisan Nyay Yojana” only supports sugarcane farmers. **(False)**
8. The “Atal Krishi Jyoti Yojana” in Gujarat aims to provide solar-powered irrigation pumps to farmers. **(True)**
9. The “Magel Tyala Shettale” scheme in Maharashtra focuses on providing rainwater harvesting facilities. **(True)**
10. The “Uzhavan App” was developed by the Kerala government for fishery-related data collection. **(False)**

27

Schemes and Programs by Private Organizations (NGOs) Involved in Promotion of Farming-Based Livelihood Opportunities

There are numerous private organizations and non-governmental organizations (NGOs) around the world that are dedicated to promoting farming-based livelihood opportunities. These organizations often implement a variety of schemes and programs aimed at improving agricultural productivity, enhancing income, and providing support to farmers. Here are some common types of schemes and programs that such NGOs might engage in:

1. Training and Capacity Building

- **Workshops and Training Programs:** NGOs often organize training sessions for farmers on modern agricultural practices, sustainable farming techniques, pest management, organic farming, and efficient water management.
- **Farmer Field Schools:** Practical learning environments where farmers can learn through hands-on experience and observation.

2. Access to Credit and Financial Services

- **Microfinance and Cooperative Societies:** NGOs may provide microloans or form cooperatives to give farmers access to credit and financial services without exorbitant interest rates.
- **Savings Groups:** Encouraging farmers to form savings groups to manage their finances better and prepare for lean seasons.

3. Market Linkages

- **Market Access Programs:** Facilitating connections between farmers and markets, including support for local and fair trade initiatives.
- **E-commerce Platforms:** Helping farmers reach broader markets via online platforms, enabling them to sell directly to consumers.

4. Input Support

- **Seed and Fertilizer Distribution:** Providing access to quality seeds and fertilizers at subsidized rates or through cooperative models.
- **Equipment Rental Services:** Offering machinery and tools for rent to reduce the financial burden of farmers.

5. Sustainable Agriculture Initiatives

- **Organic Farming Promotion:** Training farmers in organic farming practices to promote environmental sustainability and higher market prices.
- **Agroecology Programs:** Encouraging diverse farming systems that enhance biodiversity and resilience.

6. Research and Innovation

- **Community-Based Research:** Involving farmers in research initiatives to develop new farming techniques adapted to local conditions.
- **Pilot Projects:** Implementing new technologies or practices in demo plots to showcase their efficacy.

7. Social Protection and Support Systems

- **Insurance Programs:** Offering crop insurance to protect farmers from losses due to natural disasters or market fluctuations.
- **Nutrition and Food Security Initiatives:** Programs aimed at improving food security, nutrition education, and awareness about healthy eating.

8. Women and Youth Empowerment Programs

- **Women's Cooperatives:** Programs focused on empowering women in agriculture, providing them with resources and training.
- **Youth Engagement Initiatives:** Encouraging youth participation in agriculture through technology and innovation, making agriculture an attractive profession.

9. Climate Resilience Programs

- **Adaptation Strategies:** Training farmers on climate-smart practices that help them adapt to changing weather patterns.
- **Water Management Programs:** Supporting activities for efficient irrigation practices, rainwater harvesting, and drought mitigation.

6. **Which private organization works towards improving climate-resilient agriculture in India?**

a) Oxfam India b) WHO

c) UNDP d) Asian Development Bank

7. **The MS Swaminathan Research Foundation (MSSRF) primarily focuses on:**

a) Large-scale farming

b) Sustainable agriculture and food security

c) Developing hybrid seeds

d) Providing only financial aid

8. **Which NGO is involved in providing digital solutions for small farmers in India?**

a) Digital Green b) Greenpeace

c) UNICEF d) ICAR

9. **What is the primary focus of Tata Trusts' agricultural programs?**

a) Developing GM crops

b) Strengthening rural economies through sustainable agriculture

c) Providing fertilizer subsidies

d) Promoting only organic farming

10. **Which of the following NGOs actively supports farmer producer organizations (FPOs) in India?**

a) World Wildlife Fund

b) Small Farmers Agribusiness Consortium (SFAC)

c) NITI Aayog

d) CRY

True/False Questions

1. PRADAN works primarily with large-scale commercial farmers. **(False)**
2. The "WADI" project by BAIF focuses on agroforestry and livelihood enhancement for tribal farmers. **(True)**
3. SEWA helps women farmers by organizing them into self-help groups and cooperatives. **(True)**

4. ICICI Foundation's "Sustainable Livelihoods Initiative" supports urban agriculture only. **(False)**
5. The Art of Living Foundation promotes "Zero Budget Natural Farming" as a cost-effective farming method. **(True)**
6. Digital Green helps farmers by providing access to digital agricultural knowledge and training. **(True)**
7. Tata Trusts only focus on dairy farming under their agricultural programs. **(False)**
8. The MS Swaminathan Research Foundation (MSSRF) promotes climate-smart agriculture and food security. **(True)**
9. Oxfam India works to promote large-scale mechanized farming. **(False)**
10. Small Farmers Agribusiness Consortium (SFAC) supports the formation and strengthening of Farmer Producer Organizations (FPOs). **(True)**

28

Role of Farming-Based Livelihood Enterprises in 21st Century in View of Circular Economy & Green Economy

Farming-based livelihood enterprises play a crucial role in promoting sustainable development, especially in the context of the circular economy and green economy. Both these economic models emphasize sustainability, resource efficiency, and environmental responsibility. Here's how farming-based livelihood enterprises contribute to these concepts:

1. Resource Efficiency

- **Waste Minimization**: Farming enterprises often utilize organic waste, such as crop residues and livestock manure, in composting or biogas production, converting waste into valuable resources and reducing landfill use.
- **Water Management**: Practices like rainwater harvesting and drip irrigation optimize water use, contributing to a more efficient agricultural model that aligns with circular economy principles.

2. Biodiversity and Ecosystem Services

- **Agroecology**: By promoting biodiversity through polyculture and intercropping, farming enterprises enhance ecosystem resilience, improve soil health, and provide habitats for various species, supporting both local ecosystems and the green economy.
- **Sustainable Practices**: Organic farming and regenerative agriculture increase carbon sequestration in soils, thereby contributing to climate mitigation efforts.

3. Local Economies and Community Resilience

- **Job Creation**: Farming-based enterprises provide employment opportunities in rural areas, helping to build resilient local economies. This is essential for reducing poverty and promoting community development in line with the circular economy.

- **Value Chains**: Localizing supply chains by connecting farmers directly with consumers (e.g., through farmer's markets or community-supported agriculture) reduces transportation emissions and creates more equitable economic opportunities.

4. Innovation and Technology

- **Sustainable Practices**: Adoption of innovative technologies and practices, such as precision farming, agroforestry, and integrated pest management, enhances productivity while minimizing environmental impact.
- **Circular Business Models**: Enterprises can adopt circular models by rethinking product life cycles, focusing on re-use, recycling, and upcycling agricultural products.

5. Education and Awareness

- **Capacity Building**: Farming-based enterprises often engage in educating farmers and consumers about sustainable practices, supporting the transition to a greener economy.
- **Community Engagement**: By fostering community involvement in sustainable practices, these enterprises help to build awareness and advocate for environmentally friendly policies and initiatives.

6. Climate Resilience

- **Adaptation Strategies**: Sustainable farming practices help build resilience to climate change by enhancing soil health, conserving water, and promoting agro-biodiversity, thereby ensuring long-term agricultural productivity.
- **Carbon Footprint Reduction**: By implementing low-carbon practices and renewable energy sources, farming enterprises can significantly reduce their carbon footprint, aligning with green economy goals.

7. Sustainable Food Systems

- **Food Security**: Farming-based enterprises contribute to local food production, reducing dependence on imported foods and increasing food sovereignty.
- **Nutrition and Health**: By focusing on diverse crop production and reducing chemical inputs, these enterprises can improve nutrition and health outcomes for communities.

Conclusion

In the 21st century, the integration of farming-based livelihood enterprises within the circular and green economic frameworks is essential to address pressing global challenges like climate change, biodiversity loss, and food security. By adopting sustainable practices, these enterprises not only protect the environment but also enhance economic resilience and social equity, making them vital players in the transition to a more sustainable future.

QUESTION BANK

Short Answer Questions

1. What is the circular economy, and how does it relate to farming-based livelihood enterprises?
2. How can green economy principles be integrated into agriculture?
3. Name two farming enterprises that promote sustainability in the circular economy.
4. What is the role of agroforestry in a green economy?
5. How does organic farming contribute to a circular economy?
6. Name a government policy that promotes sustainable farming enterprises in the 21st century.
7. What is the significance of waste-to-value farming techniques in a circular economy?
8. How can precision agriculture help in achieving sustainability goals?
9. Name one successful example of a farming-based circular economy model in India.
10. How do Farmer Producer Organizations (FPOs) support green economic practices in agriculture?

Multiple Choice Questions (MCQs)

1. What is the main goal of the circular economy in agriculture?

a) Maximizing waste production

b) Promoting linear economic models

c) Reducing resource waste through recycling and regeneration

d) Increasing chemical fertilizer use

2. Which of the following is a key principle of the green economy in farming?

a) Overuse of groundwater

b) Reducing carbon emissions and enhancing biodiversity

c) Relying on monoculture farming

d) Encouraging excessive pesticide use

3. **How does integrated farming contribute to a circular economy?**
   a) By creating wasteful production cycles
   b) By reducing resource efficiency
   c) By utilizing by-products and minimizing waste
   d) By encouraging the use of synthetic chemicals
4. **Which of the following best describes a sustainable farming enterprise?**
   a) One that uses excessive synthetic inputs
   b) One that maximizes yield at any environmental cost
   c) One that adopts eco-friendly techniques and conserves resources
   d) One that focuses only on profit generation
5. **Which farming technique aligns best with the principles of the green economy?**
   a) Slash-and-burn farming
   b) Organic and regenerative agriculture
   c) Heavy pesticide use in monoculture
   d) Single-season cropping with high chemical inputs
6. **What is one major benefit of a circular economy approach in agriculture?**
   a) Increased dependency on fossil fuels
   b) Reduced soil fertility
   c) Enhanced resource efficiency and waste minimization
   d) Increased pollution levels
7. **What role does composting play in the circular economy?**
   a) It increases chemical fertilizer dependence
   b) It reduces soil health
   c) It recycles organic waste into valuable nutrients
   d) It leads to higher carbon emissions
8. **How does precision farming support green economic principles?**
   a) By optimizing resource use and reducing waste
   b) By increasing excessive water consumption
   c) By promoting overuse of fertilizers
   d) By encouraging monoculture farming

9. **Which of the following is an example of a farming-based livelihood enterprise that follows circular economy principles?**
   a) Conventional monoculture farming
   b) Vermicomposting and biofertilizer production
   c) Heavy mechanized farming without crop rotation
   d) Overuse of groundwater for irrigation

10. **Why is agroforestry considered a sustainable farming approach in a green economy?**
   a) It promotes deforestation
   b) It integrates trees with crops and livestock to enhance biodiversity
   c) It eliminates biodiversity by planting only one species
   d) It increases dependency on chemical pesticides

True/False Questions

1. The circular economy in farming focuses on maximizing waste and pollution. **(False)**
2. Green economy principles encourage sustainable and eco-friendly farming practices. **(True)**
3. Organic farming is not a part of the circular economy. **(False)**
4. Precision agriculture helps in reducing the excessive use of water and fertilizers. **(True)**
5. Composting agricultural waste supports the circular economy by recycling nutrients back into the soil. **(True)**
6. Monoculture farming is the most sustainable approach for a green economy. **(False)**
7. Integrated farming systems help in achieving zero-waste farming models. **(True)**
8. The use of renewable energy sources like solar-powered irrigation is aligned with green economy principles. **(True)**
9. Chemical-intensive farming is the best way to achieve a sustainable farming-based livelihood. **(False)**
10. Agroforestry helps enhance biodiversity while supporting farming-based livelihood enterprises. **(True)**

29

Role of Farming-Based Livelihood Enterprises in 21st Century in View of Climate Change, Digitalization and Changing Lifestyle

Farming-based livelihood enterprises are evolving in the 21st century, significantly influenced by climate change, digitalization, and changing lifestyles. These factors converge to reshape agricultural practices, economic viability, and rural community development. Here's an in-depth analysis of their roles in this dynamic landscape:

1. Adapting to Climate Change

- **Resilience Building**: Farming enterprises are increasingly focused on developing resilience against climate change impacts. This includes adopting sustainable farming practices, such as crop rotation, agroforestry, and organic farming, which can enhance soil health and biodiversity.
- **Innovative Crop Choices**: Farmers are diversifying crop choices to include climate-resilient varieties that can tolerate heat, drought, and flooding. This helps secure food supply chains and farmers' income against climate variability.
- **Water Management**: Efficient water management systems, such as rainwater harvesting and drip irrigation, are being implemented to mitigate water scarcity and improve irrigation efficiency.

2. Embracing Digitalization

- **Precision Agriculture**: The advent of technology has introduced precision agriculture, enabling farmers to utilize sensors, drones, and satellite imagery. This technology helps optimize inputs (like seeds, fertilizer, and water), leading to better yields and reduced environmental impact.

- **Data-Driven Decisions**: Digital platforms allow farmers to access real-time data on weather patterns, market prices, and agricultural trends. This information empowers them to make informed decisions, improving both productivity and profitability.
- **E-Commerce and Market Access**: Digital platforms facilitate direct access to markets, enabling farmers to sell their products online. This reduces dependency on intermediaries, increases profit margins, and broadens market reach.

3. Responding to Changing Lifestyles

- **Health-Conscious Consumers**: With growing awareness of nutritional health, there is a rising demand for organic and locally sourced food. This trend encourages farmers to adopt organic practices and engage in community-supported agriculture (CSA) programs, Livelihood enterprises that provide fresh produce to local consumers.
- **Agri-Tourism**: Incorporating tourism into farming enterprises can provide additional income streams. Farms offering educational tours, farm-to-table experiences, or recreational activities can attract urban visitors looking for authentic and sustainable experiences.
- **Value Addition and Processing**: As consumers seek convenience and variety, farming enterprises can engage in value-added processing (e.g., making jams, sauces, or dried fruit). This not only diversifies income but also creates local jobs and reduces food waste.

4. Empowering Communities

- **Job Creation**: Farming-based livelihood enterprises can create jobs in rural areas, helping to curb urban migration and sustain local economies. This includes direct employment in farming, processing, logistics, and retail sectors.
- **Skill Development**: Training programs that focus on sustainable practices, digital skills, and entrepreneurship can empower farmers and their families, enhancing community resilience and economic opportunity.
- **Collaboration and Cooperatives**: By forming cooperatives, farmers can pool resources, share knowledge, and gain better bargaining power. This collective approach can enhance local economies and foster innovation.

Conclusion

In the 21st century, farming-based livelihood enterprises play a pivotal role in addressing the challenges posed by climate change, leveraging digitalization, and adapting to changing lifestyles. By embracing sustainability, technology, and community engagement, these enterprises not only ensure food security but also promote economic stability and resilience in rural areas. The future of agriculture lies in its ability to innovate and adapt, transforming traditional practices to meet the demands of a rapidly changing world.

QUESTION BANK

Short Answer Questions

1. How does climate change affect farming-based livelihood enterprises?
2. What role does digitalization play in modern agricultural practices?
3. Name two climate-smart farming techniques that help mitigate the impact of climate change.
4. How do changing consumer lifestyles influence agricultural production?
5. What is precision agriculture, and how does it benefit farmers?
6. How can artificial intelligence (AI) be used in agriculture?
7. What are some examples of climate-resilient crops suitable for sustainable farming?
8. How does e-commerce contribute to farming-based livelihood enterprises?
9. Name one mobile app that helps farmers with weather forecasting and market access.
10. What is the significance of hydroponics and vertical farming in modern agriculture?

Multiple Choice Questions (MCQs)

1. Which of the following farming practices is most suitable for adapting to climate change?

a) Monoculture farming
b) Climate-smart agriculture
c) Slash-and-burn farming
d) Excessive pesticide use

2. Which technology helps farmers monitor soil health and water usage?

a) Blockchain
b) Drones and IoT sensors
c) Radio broadcasting
d) Paper-based record-keeping

3. **Which of the following best describes the impact of digitalization on farming?**
    a) Reduced access to markets
    b) Increased productivity and efficiency
    c) Increased reliance on manual labor
    d) Less access to real-time information
4. **What is the role of AI in modern agriculture?**
    a) Increasing human labor dependency
    b) Predicting weather patterns and optimizing farm operations
    c) Promoting traditional plowing techniques
    d) Reducing access to smart irrigation systems
5. **Which of the following is an example of digitalization in agriculture?**
    a) Manual irrigation systems
    b) Use of AI-driven pest monitoring systems
    c) Traditional crop rotation without data analysis
    d) Paper-based farm record-keeping
6. **What is one major advantage of vertical farming in urban areas?**
    a) It increases water consumption
    b) It reduces land use while improving yield efficiency
    c) It requires large open fields
    d) It has no impact on modern agriculture
7. **Which factor is influencing the shift toward organic and natural farming?**
    a) Increased use of chemical fertilizers
    b) Growing consumer demand for healthy and sustainable food
    c) Limited interest in environmental conservation
    d) Reduced awareness of climate change impacts
8. **Which of the following platforms helps farmers sell their products directly to consumers?**
    a) WhatsApp
    b) Amazon, BigBasket, and Ninjacart
    c) Printed brochures
    d) Conventional supply chains without digital integration

9. **What is one way digital farming helps reduce greenhouse gas emissions?**
   a) By promoting excessive irrigation
   b) By using AI-driven precision farming techniques
   c) By discouraging organic farming
   d) By increasing the use of synthetic pesticides

10. **What impact has changing consumer lifestyles had on farming enterprises?**
    a) Reduced demand for organic products
    b) Increased demand for processed and fast foods only
    c) Greater focus on sustainability, organic products, and farm-to-table supply chains
    d) No change in farming methods

True/False Questions

1. Climate change has no impact on farming-based livelihood enterprises. **(False)**
2. AI and IoT technology can help farmers optimize water usage and improve crop productivity. **(True)**
3. Digitalization in agriculture leads to reduced farm efficiency and productivity. **(False)**
4. Hydroponics and vertical farming are emerging solutions for urban agriculture. **(True)**
5. Changing lifestyles have decreased the demand for organic and sustainable agricultural products. **(False)**
6. Climate-smart agriculture aims to improve sustainability and reduce environmental impact. **(True)**
7. Smart irrigation systems waste more water than traditional irrigation methods. **(False)**
8. E-commerce platforms like Amazon and Ninjacart enable farmers to sell directly to consumers. **(True)**
9. Blockchain technology has no role in improving traceability and transparency in the agricultural supply chain. **(False)**
10. Drones can be used for precision farming, including crop monitoring and spraying pesticides. **(True)**.

Practical (Experiments)

1

Survey of Farming Systems and Agricultural-Based Livelihood Enterprises

A survey of farming systems and agricultural-based livelihood enterprises is a research method used to gather information about the different ways farmers organize their agricultural activities and how these activities contribute to their livelihoods. It aims to provide a comprehensive understanding of the diverse farming practices, resource management strategies, and income-generating activities within a specific region or community.

Objectives

Identify and classify farming systems: This involves understanding the different types of farming practices prevalent in the area, such as crop-livestock systems, mixed farming, specialized cropping, agroforestry, etc. Classification may be based on factors like scale, technology used, crops grown, livestock raised, and integration with other enterprises.

Assess resource use and management: This includes examining how farmers utilize natural resources like land, water, soil, and biodiversity. It also covers management practices like irrigation, fertilization, pest control, and soil conservation.

Analyze livelihood strategies: The survey investigates how agriculture contributes to the overall livelihoods of farming households. This includes income from crop and livestock production, as well as off-farm income sources.

Evaluate the economic viability of agricultural enterprises: This involves assessing the profitability and sustainability of different farming activities. It may include analyzing production costs, market access, and value addition opportunities.

Identify constraints and opportunities: The survey aims to pinpoint the challenges faced by farmers, such as access to credit, markets, technology, and information. It also seeks to identify potential opportunities for improving agricultural productivity and livelihoods.

Inform policy and development interventions: The survey results can be used to inform the design and implementation of agricultural policies, development programs, and extension services.

Scope

The scope of the survey can vary depending on the research objectives and available resources. It may cover:

Geographic area: A specific village, region, or country.

Target population: Smallholder farmers, commercial farmers, specific ethnic groups, or farmers engaged in particular agricultural activities.

Farming systems: All farming systems in the area or focus on specific types of systems.

Livelihood enterprises: Agriculture-related businesses, such as processing, marketing, and input supply.

Methodology

A combination of quantitative and qualitative methods is typically used:

Household surveys: Structured questionnaires are administered to a representative sample of farming households to collect data on farming practices, resource use, income, and other relevant indicators. Sampling techniques like random sampling, stratified sampling, or cluster sampling are used to ensure representativeness.

Key informant interviews: Interviews with local leaders, extension officers, agricultural experts, and other knowledgeable individuals to gather insights into farming systems, challenges, and opportunities in the area.

Focus group discussions: Group discussions with farmers to explore specific topics in more detail and to capture diverse perspectives.

Field observations: Direct observation of farming practices, resource conditions, and market activities to complement the data collected through surveys and interviews.

Secondary data review: Review of existing reports, statistics, and other relevant documents to provide context and background information.

Data Collection

Questionnaire Design: The design of the questionnaire is crucial for collecting accurate and reliable data. It should be clear, concise, and relevant to the research objectives. The questionnaire typically includes sections on:

- Household demographics

- Land ownership and use
- Crop production (area, yields, inputs, costs, prices)
- Livestock production (types, numbers, inputs, costs, prices)
- Resource management practices (irrigation, fertilization, pest control)
- Income sources (farm and off-farm)
- Access to credit, markets, and information
- Constraints and challenges

Training of enumerators: Thorough training of enumerators is essential to ensure that they understand the questionnaire and can administer it accurately.

Data entry and cleaning: The collected data is entered into a database and cleaned to identify and correct any errors.

Data Analysis

Descriptive statistics: Calculation of summary statistics, such as means, percentages, and frequencies, to describe the characteristics of the sample.

Cross-tabulations: Analysis of relationships between different variables, such as farming system type and income level.

Regression analysis: Statistical modeling to identify factors that influence agricultural productivity and livelihoods.

Qualitative data analysis: Analysis of interview transcripts and field notes to identify key themes and patterns.

Spatial analysis: Use of GIS software to map farming systems, resource use patterns, and other spatial data.

Reporting

The survey findings are typically presented in a comprehensive report that includes:

Executive summary: A brief overview of the key findings and recommendations.

2

Study of Components of Important Farming-Based Livelihood Models/Systems in Different Agro-Climatic Zones

Introduction

Background information on the study area and research objectives.

Methodology: Description of the data collection and analysis methods.

Results: Presentation of the survey findings, including tables, figures, and maps.

Discussion: Interpretation of the findings and their implications.

Conclusions and recommendations: Summary of the key conclusions and recommendations for policy and development interventions.

Example Questions for a Farming Systems Survey

General Household Information

What is the size of your household?

What is the highest level of education attained by the household head?

What is the primary occupation of the household head?

Land Use

How much lands do you own/rent/cultivate?

What crops did you grow on your land in the last year?

What is the area allocated to each crop?

Crop Production

What were your yields for each crop?

What inputs did you use (fertilizers, pesticides, seeds)?

What were your costs of production for each crop?

Where do you sell your crops (market, middleman, etc.)?

What prices did you receive for your crops?

Livestock Production

What types of livestock do you raise?

How many animals of each type do you have?

What inputs do you use for livestock production (feed, veterinary services)?

What were your costs of production for livestock?

Where do you sell your livestock products (milk, meat, eggs)?

What prices did you receive for your livestock products?

Resource Management

Do you irrigate your crops? If so, what is the source of water?

What types of fertilizers do you use?

What methods do you use to control pests and diseases?

Do you practice soil conservation measures?

Income and Livelihood:

What is your total household income from agriculture?

What are your other sources of income (off-farm employment, remittances)?

What are your major expenses?

Access to Services:

Do you have access to credit?

Do you have access to agricultural extension services?

Do you have access to market information?

Challenges and Opportunities

What are the major challenges you face in farming?

What are the opportunities for improving your agricultural productivity and livelihoods?

By collecting and analyzing this type of information, researchers and policymakers can gain a better understanding of farming systems and agricultural-based livelihoods and develop effective strategies to promote sustainable agricultural development and improve the well-being of farming communities.

3

Study of Components of Important Farming- Based Livelihood Models/ Systems in Different Agro-Climatic Zones

The study of components of important farming-based livelihood models/ systems in different agro-climatic zones is crucial to understand the diversity of agricultural practices and their impact on rural livelihoods. Here's a breakdown of the components and farming systems in different agro-climatic zones:

Agro-climatic zones

1. Tropical zone: Characterized by high temperatures and high rainfall. Crops such as rice, maize, and sugarcane are commonly grown.
2. Sub-tropical zone: Has a moderate climate with warm summers and cool winters. Crops like wheat, barley, and vegetables are widely cultivated.
3. Temperate zone: Experiences cold winters and warm summers. Crops such as wheat, oats, and fruits are commonly grown.
4. Arid zone: Characterized by low rainfall and high temperatures. Crops like millet, sorghum, and cotton are grown.

Components of farming-based livelihood models/systems

1. Crops: Various crops are grown in different agro-climatic zones, including staples like rice, wheat, and maize, as well as cash crops like cotton, tobacco, and sugarcane.
2. Livestock: Animals like cattle, buffalo, sheep, and goats are raised for milk, meat, and draft purposes.
3. Irrigation: Water management systems, including canals, wells, and groundwater extraction, support crop growth in areas with inadequate rainfall.

4. Farm machinery: Tractors, plows, and other equipment are used to mechanize farming operations, increasing efficiency and reducing labour.
5. Organic and integrated farming: Practices like crop rotation, composting, and integrated pest management are used to maintain soil fertility and reduce chemical usage.
6. Value chain development: Farmers engage in processing, packaging, and marketing their products to increase their income and competitiveness.
7. Credit and insurance: Access to credit and insurance helps farmers manage risks, invest in their farms, and improve their livelihoods.

Farming systems in different agro-climatic zones

1. Tropical zone

- Rice-based systems: Rice is the primary crop, often combined with fish farming or livestock.
- Coconut-based systems: Coconut is the main crop, with intercrops like spices, fruits, or vegetables.

2. Sub-tropical zone

- Wheat-based systems: Wheat is the primary crop, often rotated with legumes or oilseeds.
- Vegetable-based systems: A variety of vegetables is grown, often using irrigation and protected cultivation.

3. Temperate zone

- Wheat-based systems: Wheat is the primary crop, often combined with livestock or dairy farming.
- Fruit-based systems: Fruits like apples, grapes, or berries are grown, often using irrigation and integrated pest management.

4. Arid zone

- Millet-based systems: Millet is the primary crop, often combined with livestock or agroforestry.
- Cotton-based systems: Cotton is the main crop, often grown using irrigation and mechanized farming.

Case studies

1. India's Punjab region: A tropical zone with a rice-wheat cropping system, where farmers use irrigation, mechanization, and credit to improve their livelihoods.

2. China's Yangtze River Delta: A sub-tropical zone with a vegetable-based system, where farmers use protected cultivation, irrigation, and value chain development to increase their income.
3. United States' Midwest region: A temperate zone with a corn-soybean cropping system, where farmers use mechanization, irrigation, and precision agriculture to optimize yields and reduce costs.
4. Australia's Murray-Darling Basin: An arid zone with a cotton-based system, where farmers use irrigation, mechanization, and water management to maintain their livelihoods.
5. These examples illustrate the diversity of farming-based livelihood models/systems in different agro-climatic zones. Understanding these components and systems can help policymakers, researchers, and development practitioners design and implement effective interventions to improve rural livelihoods and promote sustainable agriculture.

4

Study of Production and Profitability of Crop Based, Livestock Based, Processing Based Integrated Farming-Based Livelihood Models

I. Crop-Based Livelihood Models

- Description: Focuses primarily on cultivating and selling crops.
- Production Analysis:
    - Crop Selection
        - Identify the specific crops grown (e.g., rice, wheat, maize, vegetables, fruits).
        - Analyse the rationale behind crop selection (market demand, suitability to local climate and soil, tradition, etc.).
        - Assess the use of crop rotation, intercropping, and cover cropping practices.
    - Agronomic Practices
        - Evaluate the efficiency and effectiveness of tillage, planting, irrigation, fertilization, and pest/disease management practices.
        - Determine the level of mechanization used in crop production.
        - Investigate the use of integrated pest management (IPM) and sustainable agricultural practices.
    - Yield
        - Measure crop yield (e.g., tons per hectare, bushels per acre).
        - Compare yields to regional and national averages.
        - Identify factors that limit yield (e.g., water stress, nutrient deficiencies, and pest /disease outbreaks).
    - Input Costs
        - Quantify the costs associated with seeds, fertilizers, pesticides, irrigation, labour, and machinery.

        - Analyse the cost-effectiveness of different inputs.
        - Examine access to credit and subsidies for inputs.
    - Labour Requirements
        - Determine the amount of labour required for different stages of crop production.
        - Assess the availability and cost of labour (family labour vs. hired labour).
        - Analyse the impact of labour shortages on production.
- Profitability Analysis
    - Gross Revenue
        - Calculate the total revenue generated from crop sales.
        - Consider different marketing channels (e.g., direct sales, wholesale markets, cooperatives).
    - Production Costs
        - Summarize all costs associated with crop production (from input costs).
    - Gross Profit
        - Calculate gross profit by subtracting total production costs from gross revenue.
        - Gross Profit=Gross Revenue−Production Costs Gross \ Profit = Gross \ Revenue - Production \ Costs Gross Profit=Gross Revenue−Production Costs
    - Net Profit
        - Consider overhead costs (e.g., land rent, insurance, taxes, depreciation of equipment).
        - Calculate net profit by subtracting overhead costs from gross profit.
        - Net Profit=Gross Profit−Overhead CostsNet \ Profit = Gross \ Profit - Overhead \ CostsNet Profit=Gross Profit−Overhead Costs
    - Profit Margins
        - Calculate profit margins (e.g., gross profit margin, net profit margin) to assess the profitability of crop production.
        - Gross Profit Margin=(Gross Profit/Gross Revenue)* 100Gross \ Profit \ Margin = (Gross \ Profit / Gross \ Revenue) * 100Gross Profit Margin=(Gross Profit/Gross Revenue)* 100

        - Net Profit Margin=(Net Profit/Gross Revenue)* 100 Net\ Profit \ Margin = (Net \ Profit / Gross \ Revenue) * 100 Net Profit Margin=(Net Profit/Gross Revenue)* 100
    - Break-Even Analysis
        - Determine the yield or price required to break even (i.e., cover all costs).
    - Risk Analysis
        - Assess the risks associated with crop production (e.g., weather-related risks, price volatility, pest/disease outbreaks).
        - Evaluate risk management strategies (e.g., crop insurance, diversification).
- II. Livestock-Based Livelihood Models
- Description: Focuses primarily on raising and selling livestock or livestock products (e.g., milk, meat, eggs, wool).
- Production Analysis:
    - Livestock Species and Breeds:
        - Identify the specific livestock species raised (e.g., cattle, goats, sheep, poultry, pigs).
        - Analyze the breeds used and their suitability to the local environment and market demand.
    - Feeding and Nutrition:
        - Evaluate feeding practices (e.g., grazing, supplementation with concentrates).
        - Assess the quality and availability of feed resources.
        - Analyze the costs associated with feed.
    - Animal Health Management
        - Examine disease prevention and control measures (e.g., vaccination, parasite control).
        - Assess the availability and cost of veterinary services.
    - Housing and Infrastructure
        - Evaluate the adequacy of housing facilities for livestock.
        - Assess the availability of water and other essential infrastructure.
    - Reproduction
        - Analyze reproductive efficiency (e.g., calving rate, lambing rate, conception rate).
        - Assess the use of artificial insemination and other reproductive technologies.

  - Production Parameters
    - Measure key production parameters such as milk yield, meat production, egg production, wool production, and growth rates.
    - Compare these parameters to regional and national averages.
- Profitability Analysis
  - Gross Revenue
    - Calculate the total revenue generated from the sale of livestock and livestock products.
    - Consider different marketing channels (e.g., direct sales, livestock auctions, processing plants).
  - Production Costs
    - Quantify the costs associated with feed, veterinary services, housing, labor, and other inputs.
  - Gross Profit:
    - Calculate gross profit by subtracting total production costs from gross revenue.
    - Gross Profit=Gross Revenue−Production CostsGross \ Profit = Gross \ Revenue - Production \ CostsGross Profit=Gross Revenue−Production Costs
  - Net Profit
    - Consider overhead costs (e.g., land rent, insurance, taxes, and depreciation of equipment).
    - Calculate net profit by subtracting overhead costs from gross profit.
    - Net Profit=Gross Profit−Overhead CostsNet \ Profit = Gross \ Profit - Overhead \ CostsNet Profit=Gross Profit−Overhead Costs
  - Profit Margins
    - Calculate profit margins (e.g., gross profit margin, net profit margin) to assess the profitability of livestock production.
    - Gross Profit Margin=(Gross Profit/Gross Revenue)*100Gross \ Profit \ Margin = (Gross \ Profit / Gross \ Revenue) * 100Gross Profit Margin=(Gross Profit/Gross Revenue)*100
    - Net Profit Margin=(Net Profit/Gross Revenue)*100Net \ Profit \ Margin = (Net \ Profit / Gross \ Revenue) * 100Net Profit Margin=(Net Profit/Gross Revenue)*100

- Break-Even Analysis
    - Determine the production level or price required to break even (i.e., cover all costs).
- Risk Analysis
    - Assess the risks associated with livestock production (e.g., disease outbreaks, price volatility, weather-related risks).
    - Evaluate risk management strategies (e.g., insurance, diversification).

- III. Processing-Based Livelihood Models
- Description: Involves processing agricultural products (crops or livestock) to add value and increase profitability. Examples include:
    - Crop Processing: Grain milling, fruit and vegetable processing, oilseed extraction.
    - Livestock Processing: Dairy processing (cheese, yogurt), meat processing, egg processing.
- Production Analysis
    - Raw Material Sourcing:
        - Identify the sources of raw materials (e.g., own farm, local farmers, wholesale markets).
        - Assess the quality and availability of raw materials.
        - Analyze the costs associated with raw material procurement.
    - Processing Technology
        - Evaluate the processing technology used (e.g., traditional methods, modern equipment).
        - Assess the efficiency and effectiveness of the processing technology.
        - Consider the energy requirements and environmental impact of the processing technology.
    - Production Capacity
        - Determine the processing capacity of the facility.
        - Analyze the utilization rate of the processing capacity.
    - Quality Control
        - Assess the quality control measures in place to ensure product safety and quality.
        - Evaluate compliance with food safety regulations.

- Packaging and Labeling
    - Examine the packaging and labeling practices used to preserve product quality and attract consumers.
- Profitability Analysis
  - Gross Revenue
    - Calculate the total revenue generated from the sale of processed products.
    - Consider different marketing channels (e.g., direct sales, wholesale markets, retail stores).
- Production Costs
  - Quantify the costs associated with raw materials, processing, packaging, labor, energy, and other inputs.
  - Gross Profit
    - Calculate gross profit by subtracting total production costs from gross revenue.
    - Gross Profit=Gross Revenue−Production CostsGross \ Profit = Gross \ Revenue - Production \ CostsGross Profit=Gross Revenue−Production Costs
  - Net Profit
    - Consider overhead costs (e.g., rent, utilities, marketing, administrative expenses, loan interest).
    - Calculate net profit by subtracting overhead costs from gross profit.
    - Net Profit=Gross Profit−Overhead CostsNet \ Profit = Gross \ Profit - Overhead \ CostsNet Profit=Gross Profit−Overhead Costs
  - Value Addition
    - Calculate the value added by processing (i.e., the difference between the value of the processed product and the value of the raw materials).
  - Profit Margins
    - Calculate profit margins (e.g., gross profit margin, net profit margin) to assess the profitability of processing.
    - Gross Profit Margin=(Gross Profit/Gross Revenue)*100Gross \ Profit \ Margin = (Gross \ Profit / Gross \ Revenue) * 100Gross Profit Margin=(Gross Profit/Gross Revenue)*100

        - Net Profit Margin=(Net Profit/Gross Revenue)□100Net \ Profit \ Margin = (Net \ Profit / Gross \ Revenue) * 100Net Profit Margin=(Net Profit/Gross Revenue)□100
    - Break-Even Analysis
        - Determine the production volume or price required to break even (i.e., cover all costs).
    - Market Analysis
        - Analyze the market demand for processed products.
        - Assess the competitive landscape.
        - Identify opportunities for product differentiation and market expansion.
- IV. Integrated Farming Systems (IFS) / Integrated Farming-Based Livelihood Models
- Description: Combines crop production, livestock rearing, and potentially other activities (e.g., aquaculture, agroforestry) on the same farm to enhance resource utilization, reduce waste, and increase overall productivity and profitability.
- Production Analysis
    - System Components
        - Identify the specific crop, livestock, and other components integrated into the farming system.
        - Analyze the interactions and synergies between the different components.
    - Resource Cycling
        - Evaluate the flow of nutrients, energy, and water within the system.
        - Assess the efficiency of resource utilization.
        - Determine the extent to which waste products from one component are used as inputs for another component (e.g., using livestock manure as fertilizer for crops).
    - Crop-Livestock Integration
        - Analyze the integration of crop and livestock production (e.g., using crop residues as livestock feed, using livestock to control weeds in crops).
    - Diversification

- Assess the level of diversification within the farming system.
- Evaluate the impact of diversification on income stability and risk reduction.

o Sustainability

- Evaluate the environmental sustainability of the farming system.
- Consider factors such as soil health, water quality, biodiversity, and greenhouse gas emissions.

- Profitability Analysis

  o Gross Revenue

  - Calculate the total revenue generated from all components of the farming system.
  - Consider the revenue from crop sales, livestock sales, and other activities (e.g., sale of processed products, agritourism).

  o Production Costs

  - Quantify the costs associated with each component of the farming system.
  - Allocate shared costs (e.g., labor, machinery) to different components.

  o Gross Profit

  - Calculate gross profit for each component and for the entire farming system.
  - Gross Profit=Gross Revenue−Production CostsGross \ Profit = Gross \ Revenue - Production \ CostsGross Profit=Gross Revenue−Production Costs

  o Net Profit

  - Consider overhead costs (e.g., land rent, insurance, taxes).
  - Calculate net profit for each component and for the entire farming system.
  - Net Profit=Gross Profit−Overhead CostsNet \ Profit = Gross \ Profit - Overhead \ CostsNet Profit=Gross Profit−Overhead Costs

  o Profit Margins

  - Calculate profit margins (e.g., gross profit margin, net profit margin) to assess the profitability of the farming system.
  - Gross Profit Margin=(Gross Profit/Gross Revenue)□100Gross \ Profit \ Margin = (Gross \ Profit / Gross \ Revenue) *

100Gross Profit Margin=(Gross Profit/Gross Revenue)□100

- Net Profit Margin=(Net Profit/Gross Revenue)□100Net \ Profit \ Margin = (Net \ Profit / Gross \ Revenue) * 100Net Profit Margin=(Net Profit/Gross Revenue)□100

  - Resource Use Efficiency
    - Analyze the efficiency of resource use in the farming system (e.g., land use efficiency, water use efficiency, nutrient use efficiency).
  - Risk Analysis
    - Assess the risks associated with the farming system (e.g., weather-related risks, price volatility, pest/disease outbreaks).
    - Evaluate the risk-reducing benefits of diversification.
- General Considerations for all Models
- Data Collection
    - Surveys of farmers using each model
    - Interviews with key informants (extension agents, researchers, market participants)
    - Collection of secondary data (e.g., agricultural statistics, market prices)
    - On-farm trials to evaluate different practices
- Economic Indicators
    - Net Farm Income
    - Return on Investment (ROI)
    - Benefit-Cost Ratio (BCR)
    - Labor Productivity
- Socio-Economic Factors
    - Farm size
    - Household income and expenses
    - Access to credit and insurance
    - Education level
    - Social capital (networks and relationships)
- Environmental Impact
  - Soil health
  - Water quality and use
  - Biodiversity
  - Greenhouse gas emissions

- Study Design
- Comparative Analysis: Compare the production and profitability of different livelihood models.
- Case Studies: Conduct in-depth case studies of successful and unsuccessful farming enterprises.
- Statistical Analysis: Use statistical methods to analyze data and identify significant relationships between variables.

By systematically analyzing these factors, you can gain a comprehensive understanding of the production and profitability of different farming-based livelihood models and identify strategies to improve the livelihoods of farmers. Remember to tailor the specific areas of focus to the context of your study (e.g., geographic location, specific crops and livestock). Good luck!

5

Field Visit of Innovative Farming System Models

A field visit to innovative farming system models would involve observing and learning about different approaches to agriculture that aim to improve efficiency, sustainability, and profitability. Here's a breakdown of what such a visit might entail, the types of systems you might encounter, and the potential benefits:

Objectives of a Field Visit

- Practical Learning: See firsthand how innovative systems are implemented and operated.
- Knowledge Exchange: Interact with farmers, researchers, and experts who are using these systems.
- Inspiration & Idea Generation: Identify potential adaptations and applications for your own farm or region.
- Networking: Connect with other individuals and organizations interested in sustainable agriculture.
- Assessment: Evaluate the real-world performance of different technologies and practices.

Types of Innovative Farming System Models You Might Encounter

1. Integrated Farming Systems (IFS):

- Concept: Combines different agricultural enterprises (e.g., crops, livestock, aquaculture) in a way that they support each other, creating synergies and reducing waste. Often includes on-farm nutrient cycling.
- Examples
    - Rice-fish farming: Raising fish in rice paddies to control pests and provide fertilizer.
    - Crop-livestock integration: Using crop residues as feed for livestock and livestock manure as fertilizer.

- Key Features
    - Resource optimization
    - Waste reduction
    - Increased biodiversity
    - Improved soil health

2. Agroforestry

- Concept: Integrates trees and shrubs into agricultural landscapes.
- Examples
    - Alley cropping: Planting crops between rows of trees.
    - Silvopasture: Integrating trees and livestock grazing.
    - Forest gardening: A multi-layered system mimicking a natural forest ecosystem.
- Key Features
    - Soil conservation
    - Carbon sequestration
    - Habitat creation
    - Diversified income streams (timber, fruits, nuts)
    - Microclimate regulation

3. Conservation Agriculture (CA)

- Concept: A set of soil management practices that minimize soil disturbance, maintain permanent soil cover, and promote crop rotation.
- Examples
    - No-till farming: Planting crops directly into the soil without plowing.
    - Cover cropping: Planting crops specifically to protect and improve the soil.
    - Residue management: Leaving crop residues on the soil surface.
- Key Features
    - Reduced soil erosion
    - Improved soil health
    - Water conservation
    - Reduced fuel consumption
    - Carbon sequestration

4. Precision Agriculture (PA)

- Concept: Using technology to monitor and manage crops and livestock with greater precision, optimizing inputs and reducing waste.
- Examples:
  - GPS-guided machinery: For precise planting, spraying, and harvesting.
  - Variable rate application: Applying fertilizers and pesticides only where needed.
  - Remote sensing: Using drones or satellites to monitor crop health and soil conditions.
  - Sensor-based irrigation: Watering crops based on real-time soil moisture measurements.
- Key Features
  - Increased efficiency
  - Reduced input costs
  - Environmental protection
  - Improved yields
  - Data-driven decision-making

5. Vertical Farming

- Concept: Growing crops in vertically stacked layers, often indoors and using controlled environment agriculture (CEA).
- Examples
  - Hydroponics: Growing plants without soil, using nutrient-rich water solutions.
  - Aeroponics: Growing plants with roots suspended in the air, periodically sprayed with nutrient solutions.
  - Aquaponics: Integrating aquaculture (raising fish) with hydroponics.
- Key Features
  - High yields in small spaces
  - Reduced water usage
  - Year-round production
  - Pest and disease control
  - Urban agriculture potential

6. Organic Farming

- Concept: A system of agriculture that emphasizes natural inputs, biodiversity, and ecological balance.
- Examples
  - Crop rotation
  - Composting
  - Biological pest control
  - Green manure
- Key Features
  - Elimination of synthetic pesticides and fertilizers
  - Improved soil health
  - Reduced environmental impact
  - Premium prices for organic produce

7. Regenerative Agriculture

- Concept: Focuses on rebuilding soil health and biodiversity, capturing carbon, and improving the water cycle. Often overlaps with Conservation Agriculture and Organic Farming.
- Examples
  - No-till farming
  - Cover cropping
  - Composting
  - Managed grazing (holistic management)
  - Agroforestry
- Key Features
  - Soil health improvement
  - Carbon sequestration
  - Increased biodiversity
  - Water cycle enhancement
  - Resilience to climate change

What to Look For During a Field Visit

- System Design: How is the system set up and operated? What are the key components and their interactions?
- Inputs & Outputs: What resources are required (water, energy, fertilizer, labor)? What are the yields and economic returns?

- Environmental Impact: How does the system affect soil health, water quality, biodiversity, and greenhouse gas emissions?
- Social Impact: How does the system affect the livelihoods of farmers and the well-being of the community?
- Scalability & Adaptability: Can the system be scaled up or adapted to different regions and farming contexts?
- Challenges & Solutions: What are the main challenges faced by farmers using the system? What strategies are being used to overcome them?
- Data & Monitoring: How is the system being monitored and evaluated? What data are being collected?

Preparing for a Field Visit

- Research: Learn about the different farming systems in advance.
- Objectives: Define your goals for the visit. What specific questions do you want to answer?
- Questions: Prepare a list of questions to ask the farmers, researchers, or experts.
- Documentation: Bring a notebook, camera, or recording device to document your observations.
- Respect: Be respectful of the farmers' time and property.
- Example Questions to Ask:
- What motivated you to adopt this farming system?
- What are the biggest benefits you've seen?
- What are the biggest challenges?
- What advice would you give to someone considering this system?
- How has this system impacted your profitability?
- How has this system impacted your quality of life?
- What kind of data are you collecting to track your progress?
- What resources or support did you find most helpful?

By carefully planning and conducting a field visit, you can gain valuable insights into innovative farming systems and identify opportunities to improve your own agricultural practices. Remember to consider the specific context of each farm and the potential trade-offs involved in adopting different technologies and approaches.

6

Visit of Agri-based Enterprises and their Functional Aspects for Integration of Production, Processing and Distributionsectors

A visit to agri-based enterprises focusing on the integration of production, processing, and distribution offers valuable insights into how these sectors can be effectively linked to improve efficiency, reduce waste, and enhance profitability. Here's a breakdown of what such a visit might entail and the functional aspects you'd likely observe:

I. Objectives of the Visit

- **Understanding Integration:** To observe firsthand how production, processing, and distribution are integrated in a real-world agri-business.
- **Identifying Best Practices:** To identify successful strategies and technologies used in the enterprise.
- **Analyzing Functional Aspects:** To examine the key functional areas and their roles in the integrated value chain.
- **Learning About Challenges:** To understand the obstacles faced by the enterprise and how they are overcome.
- **Networking:** To establish contacts with industry professionals and learn from their experiences.

II. Agri-Based Enterprises to Consider

The type of enterprise you visit will significantly shape your observations. Here are a few examples:

- **Dairy Processing Plant:** Integrates milk production at farms with processing (pasteurization, cheese-making) and distribution to retailers.
- **Fruit and Vegetable Processing Unit:** Links farms producing fruits and vegetables to processing (canning, freezing, juice extraction) and distribution networks.

- **Grain Processing Facility:** Integrates grain production with milling, packaging, and distribution.
- **Integrated Poultry or Meat Processing:** Connects livestock farms with processing plants and distribution channels.
- **Aquaculture Farm with Processing and Retail:** Combines fish farming with on-site processing and direct sales.
- **III. Functional Aspects to Observe During the Visit:**
- Here's a breakdown of the key functional areas to examine, linking production, processing, and distribution:

A. Production (Farm Level)

- **Farming Practices**
  - **Crop Selection/Animal Husbandry:** What types of crops/animals are raised, and why? Are they chosen for processing suitability?
  - **Technology Adoption:** What technologies are used to optimize yield and quality (e.g., precision farming, automated irrigation, advanced feeding systems)?
  - **Sustainable Practices:** Are sustainable farming methods employed (e.g., crop rotation, organic farming, responsible animal welfare)?
- **Input Management**
  - **Sourcing of Inputs:** How are seeds, fertilizers, feed, and other inputs sourced? Are there contracts with suppliers?
  - **Quality Control:** How is the quality of inputs ensured?
- **Harvesting/Collection**
  - **Timing and Methods:** How is harvesting/collection timed to optimize quality and minimize losses? What methods are used (manual, mechanical)?
  - **Transportation to Processing:** How are raw materials transported to the processing facility? Are there specialized vehicles or handling procedures? What is the distance from the farm to the processing facility?
- **Farmer Relationships**
  - **Contract Farming:** Are farmers contracted to supply the enterprise? What are the terms of the contracts?
  - **Extension Services:** Does the enterprise provide extension services or support to farmers (e.g., training, technical assistance)?

- **Fair Pricing:** How are prices determined for the raw materials supplied by farmers?

B. Processing (Factory Level)

- **Raw Material Handling**
  - **Receiving and Inspection:** How are raw materials received, inspected, and graded?
  - **Storage:** How are raw materials stored to maintain quality and prevent spoilage?
- **Processing Operations**
  - **Technology and Equipment:** What technologies and equipment are used in the processing operations? Are they modern and efficient?
  - **Process Control:** How is the processing controlled to ensure consistent quality and safety?
  - **Waste Management:** How is waste generated during processing managed? Are there efforts to reduce, reuse, or recycle waste?
- **Quality Assurance**
  - **Quality Control Procedures:** What quality control procedures are in place at each stage of processing?
  - **Certifications:** Does the enterprise have any quality certifications (e.g., ISO, HACCP)?
  - 0**Traceability:** How is traceability maintained from raw materials to finished products?
- **Packaging**
  - **Packaging Materials:** What types of packaging materials are used? Are they environmentally friendly?
  - **Labeling:** How are products labeled with information about ingredients, nutrition, and shelf life?
- **Inventory Management**
  - **Storage of Finished Goods:** How are finished goods stored to maintain quality and prevent damage?
  - **Inventory Control Systems:** What inventory control systems are used to track stock levels and manage orders?

C. Distribution (Market Level)

- **Transportation**
  - **Transportation Methods:** How are finished products transported to distribution centers and retail outlets? Are specialized vehicles used (e.g., refrigerated trucks)?

  - **Logistics Management:** How is the logistics of transportation managed to ensure timely and efficient delivery?
- **Distribution Channels**
  - **Wholesalers:** Are wholesalers used to distribute products?
  - **Retailers:** What types of retailers are used (e.g., supermarkets, grocery stores, convenience stores)?
  - **Direct Sales:** Are products sold directly to consumers (e.g., farmers' markets, online sales)?
- **Marketing and Sales**
  - **Branding:** How are products branded and marketed to consumers?
  - **Pricing Strategies:** What pricing strategies are used?
  - **Promotional Activities:** What promotional activities are used to increase sales?
- **Customer Service**
  - **Handling Customer Complaints:** How are customer complaints handled?
  - **Feedback Mechanisms:** What feedback mechanisms are in place to gather customer feedback?
- **Supply Chain Management**
  - **Order Fulfillment:** How are orders fulfilled from retailers and distributors?
  - **Demand Forecasting:** How is demand forecasted to ensure adequate supply?

IV. Key Questions to Ask During the Visit

- To gain a comprehensive understanding, ask questions related to:
- **Integration Strategy**
  - What are the benefits of integrating production, processing, and distribution for this enterprise?
  - What were the main drivers for integrating these sectors?
  - How is the integration managed and coordinated?
- **Technology Adoption**
  - What technologies are used in each sector to improve efficiency and quality?
  - How has technology impacted the integration of the value chain?

- **Sustainability**
  - What sustainable practices are employed in each sector?
  - How does the enterprise minimize its environmental impact?
- **Challenges and Solutions**
  - What are the main challenges faced by the enterprise in integrating these sectors?
  - How are these challenges being addressed?
- **Financial Performance**
  - What are the key financial indicators used to measure the performance of the integrated enterprise?
  - How has integration impacted profitability and return on investment?
- **Future Plans**
  - What are the future plans for the enterprise in terms of expanding operations, adopting new technologies, or entering new markets?

V. Example Scenario: Visiting a Dairy Processing Plant

- **Production (Farm):** Observe dairy farms supplying milk. Note the breeds of cows, feeding practices, milking methods (manual vs. automated), milk storage and cooling techniques, and transportation to the plant. Inquire about contracts between the plant and farmers, quality control measures at the farm level (e.g., testing for antibiotics), and support provided to farmers (e.g., veterinary services).
- **Processing (Plant):** Observe the milk receiving process, pasteurization, homogenization, packaging (different sizes, materials), and production of other dairy products (cheese, yogurt, butter). Note the equipment used, quality control procedures at each stage, waste management practices, and energy efficiency measures.
- **Distribution:** Observe how finished products are stored and transported (refrigerated trucks), distribution channels (wholesalers, retailers, direct delivery), and marketing strategies (branding, pricing, promotions).

• VI. Post-Visit Analysis

- After the visit, analyze your observations and document your findings. Consider the following:
- **Strengths and Weaknesses:** Identify the strengths and weaknesses of the integrated enterprise.
- **Opportunities and Threats:** Identify potential opportunities for improvement and potential threats to the enterprise.

- **Lessons Learned:** Document the key lessons learned from the visit.

Recommendations: Develop recommendations for improving the integration of production, processing, and distribution in other agri-based enterprises.

By carefully observing and analyzing the functional aspects of an agri-based enterprise, you can gain valuable insights into the complexities and benefits of integrating production, processing, and distribution sectors. This knowledge can be applied to improve the efficiency, sustainability, and profitability of agri-businesses.

7

Study of Agri-Enterprises Involved in Industry Sectors (Value Chain Models)

1. What are Agri-enterprises?

- Agri-enterprises encompass businesses involved in any stage of the agricultural value chain, beyond just primary production (farming). This includes:
- Input Suppliers: Companies that provide seeds, fertilizers, pesticides, machinery, animal feed, etc., to farmers.
- Processors: Businesses that transform raw agricultural products into more refined or consumer-ready goods (e.g., milling grains, processing meat, canning fruits and vegetables, dairy processing).
- Manufacturers: Companies that use agricultural products as ingredients in other manufactured goods (e.g., food and beverage companies, textile manufacturers using cotton).
- Distributors and Retailers: Businesses that move agricultural products from producers/processors to consumers (e.g., wholesalers, supermarkets, restaurants).
- Service Providers: Companies that provide services to support the agricultural sector (e.g., transportation, storage, marketing, finance, insurance, agricultural consulting, technology providers).

2. The Value Chain Concept

The value chain describes the full range of activities required to bring a product or service from conception through the different phases of production (involving a combination of physical transformation and the input of various producer services), delivery to final consumers, and final disposal after use.

3. Value Chain Models in Agri-enterprises

Value chain analysis helps to understand how value is created and captured at each stage of the agricultural process. Here are some common models and approaches:

- Linear Value Chain: This is a simplified, sequential model that traces the steps from input supply to production, processing, distribution, and consumption. While useful for initial analysis, it doesn't always capture the complexities of real-world systems.
- Networked Value Chain: Recognizes that agricultural value chains are often complex networks involving multiple actors and relationships. This model emphasizes the importance of coordination, collaboration, and information flow among different stakeholders. It acknowledges feedback loops and iterative processes.
- Global Value Chains (GVCs): This is particularly relevant for agri-enterprises involved in international trade. GVC analysis examines how different stages of the value chain are distributed across countries, considering factors like labor costs, technology, and market access.
- Integrated Value Chains: Involve closer relationships and coordination between different actors in the chain. This could involve vertical integration (e.g., a company owning both farms and processing facilities) or horizontal integration (e.g., a cooperative of farmers).
- Supply Chain Management (SCM) Perspective: Often used interchangeably, but SCM has a stronger focus on the logistics and efficiency of moving products through the chain. Value chain analysis often takes a broader view, considering social, economic, and environmental impacts.

4. Key Areas of Study within Agri-enterprise Value Chains

- Value Addition: Identifying opportunities to increase the value of agricultural products through processing, packaging, branding, or other enhancements.
- Market Access:Analyzing how agri-enterprises can access markets (both domestic and international) and meet consumer demand.
- Competitiveness: Evaluating the competitiveness of agri-enterprises relative to others in the same market. This involves examining factors like cost structure, productivity, quality, and innovation.
- Sustainability: Assessing the environmental and social impacts of agri-enterprise value chains. This includes issues like resource use, pollution, labor practices, and food security.
- Inclusiveness: Ensuring those smallholder farmers and other marginalized groups can participate in and benefit from agri-enterprise value chains.

- Resilience: Examining the ability of agri-enterprise value chains to withstand shocks and disruptions, such as climate change, disease outbreaks, or economic crises.
- Technology Adoption:Analyzing how agri-enterprises can adopt new technologies to improve efficiency, productivity, and sustainability. This could include precision agriculture, biotechnology, or digital technologies.
- Finance and Investment: Understanding the financial needs of agri-enterprises and identifying sources of funding.

5. Analytical Tools and Methods

- Value Chain Mapping: Visually representing the different stages and actors in a value chain.
- Cost-Benefit Analysis: Evaluating the economic costs and benefits of different value chain interventions.
- SWOT Analysis: Identifying the strengths, weaknesses, opportunities, and threats facing agri-enterprises in a particular value chain.
- Porter's Five Forces:Analyzing the competitive forces shaping the industry.
- Benchmarking: Comparing the performance of agri-enterprises against best practices.
- Econometric Modeling: Using statistical methods to analyze the relationships between different variables in the value chain.
- Social Network Analysis: Mapping the relationships between different actors in the value chain to understand information flows and power dynamics.
- Life Cycle Assessment (LCA): Assessing the environmental impacts of a product or service throughout its entire life cycle.

6. Examples of Research Questions

- How can smallholder farmers be integrated into high-value agricultural supply chains?
- What are the key constraints to the development of a specific agricultural value chain in a particular region?
- How can technology be used to improve the efficiency and sustainability of agricultural value chains?
- What are the impacts of trade liberalization on agricultural value chains in developing countries?

- How can agri-enterprises build more resilient value chains in the face of climate change?

7. Challenges in Studying Agri-enterprise Value Chains

- Data Availability: Obtaining reliable data on all stages of the value chain can be challenging, especially in developing countries.
- Complexity: Agricultural value chains can be highly complex, with many different actors and relationships.
- Seasonality: Agricultural production is often seasonal, which can affect the availability of products and the dynamics of the value chain.
- Heterogeneity: Agricultural products can vary widely in quality and characteristics, which can affect processing and marketing.
- Political and Policy Context: Government policies and regulations can have a significant impact on agricultural value chains.
- Examples of Value Chain Interventions:
- Improving access to inputs: Providing farmers with access to high-quality seeds, fertilizers, and other inputs.
- Providing training and technical assistance: Helping farmers to improve their production practices and manage their businesses.
- Investing in infrastructure: Building roads, storage facilities, and other infrastructure to improve the efficiency of the value chain.
- Promoting market linkages: Connecting farmers with buyers and processors.
- Supporting the development of value-added products: Helping farmers to process and package their products to increase their value.
- Strengthening farmer organizations: Supporting the formation of farmer cooperatives and other organizations to give farmers more bargaining power.
- Improving access to finance: Providing farmers with access to credit and other financial services.

By understanding the dynamics of agri-enterprise value chains, policymakers, businesses, and researchers can identify opportunities to improve the efficiency, sustainability, and inclusiveness of the agricultural sector.

I hope this comprehensive overview is helpful. If you have any specific questions or would like me to elaborate on any of these points, please ask!

8

Study of Agri-Enterprises Involved in Service Sectors (Value Chain Models)

The study of agri-enterprises involved in service sectors within the context of value chain models is a crucial area for understanding and improving the efficiency, profitability, and sustainability of agricultural systems. Here's a breakdown of key aspects:

1. Understanding Agri-Enterprise in Service Sectors

- Definition: Agri-enterprises in service sectors are businesses that provide services that support and enhance agricultural production, processing, distribution, and marketing. These services are integral parts of the agricultural value chain.
- Examples of Service Sector Agri-Enterprises
  - Input Suppliers: Seed companies, fertilizer distributors, pesticide applicators, irrigation equipment providers.
  - Financial Services: Microfinance institutions providing agricultural loans, crop insurance providers.
  - Extension Services: Agricultural consultants, agronomists offering advice on farming practices, soil testing labs.
  - Logistics and Transportation: Trucking companies transporting produce, cold storage facilities, warehousing services.
  - Processing and Packaging: Food processing plants, packaging material suppliers, labeling services.
- Marketing and Distribution: Wholesale markets, retailers, e-commerce platforms for agricultural products, export agents.
- Technology and Information Services: Precision agriculture technology providers, farm management software companies, agricultural data analytics firms, online marketplaces.
- Certification and Quality Assurance: Organic certification bodies, food safety testing labs.

2. Value Chain Models in Agriculture

- Definition: A value chain describes the full range of activities required to bring a product or service from conception through the different phases of production (involving a combination of physical transformation and the input of various producer services), delivery to final consumers, and final disposal after use. In agriculture, it encompasses all activities from input supply, production, processing, and marketing to the consumer.
- Key Components of an Agricultural Value Chain
  - Primary Production: Farming activities.
  - Input Supply: Providing seeds, fertilizers, pesticides, equipment, etc.
  - Aggregation: Collecting produce from farmers (often done by intermediaries).
  - Processing: Transforming raw agricultural products (e.g., milling grain, canning fruits).
  - Packaging: Preparing products for distribution and sale.
  - Storage: Preserving products to extend shelf life and manage supply.
  - Transportation: Moving products between different stages of the value chain.
  - Marketing and Distribution: Selling products to consumers (wholesale, retail, export).
  - Consumption: The end user consuming the product.
- Value Chain Analysis: A process of examining the different actors and activities within a value chain to identify opportunities for improvement and value addition. This involves mapping the chain, understanding the relationships between actors, and assessing the costs and benefits at each stage.
- AnalyzingAgri-Enterprises within Value Chains:
- Focus Areas: When studying agri-enterprises within value chains, consider the following:
  - Role and Contribution: How does the agri-enterprise contribute to the overall efficiency and effectiveness of the value chain? What value do they add at their specific stage?
  - Relationships with Other Actors: What are the relationships between the agri-enterprise and other players in the value chain (farmers, processors, retailers, etc.)? Are these relationships collaborative, competitive, or exploitative?

- Market Access: How does the agri-enterprise connect farmers to markets? Do they help farmers access better prices or new markets?
- Technology Adoption: Does the agri-enterprise promote the adoption of new technologies and practices by farmers?
- Sustainability: Are the agri-enterprise's practices environmentally sustainable? Do they promote sustainable farming practices?
- Impact on Farmers' Incomes: How does the agri-enterprise affect the incomes of farmers? Do they help farmers increase their productivity and profitability?
- Challenges and Constraints: What challenges and constraints does the agri-enterprise face (e.g., access to finance, lack of infrastructure, regulatory barriers)?
- Opportunities for Improvement: What opportunities exist for the agri-enterprise to improve its performance and contribute more effectively to the value chain?

4. Value Chain Models - Examples & Considerations

- Linear Value Chains: Traditional, sequential flow of products from producers to consumers. Often characterized by limited coordination and information sharing.
- Collaborative Value Chains: Involve closer relationships between actors, with increased coordination and information sharing. Can lead to higher quality products, improved efficiency, and greater value for all participants.
- Contract Farming: Agreements between farmers and processors/buyers that specify the terms of production and sale. Provides farmers with assured markets and access to inputs, while providing buyers with a reliable supply of product.
- Integrated Value Chains: A single company controls multiple stages of the value chain (e.g., a company owns both farms and processing plants).
- Short Value Chains: Direct sales from producers to consumers (e.g., farmers' markets, community-supported agriculture). Reduces the number of intermediaries and allows farmers to capture a larger share of the value.
- Global Value Chains: Involve activities spanning multiple countries, from raw material sourcing to manufacturing and distribution.

5. Research Methodologies

- Value Chain Mapping: Visually representing the different stages and actors in the value chain.
- Surveys: Collecting data from farmers, agri-enterprises, and other stakeholders.
- Interviews: Gathering in-depth information from key informants.
- Case Studies: Examining specific examples of agri-enterprises and their role in value chains.
- Quantitative Analysis: Using statistical methods to analyze data and measure the impact of agri-enterprises on value chain performance.
- Qualitative Analysis:Analyzing qualitative data (e.g., interview transcripts, focus group discussions) to understand the perspectives and experiences of stakeholders.

6. Key Performance Indicators (KPIs)

- Value Addition: How much value is added at each stage of the chain?
- Efficiency: How efficiently are resources used (e.g., water, energy, labor)?
- Profitability: What are the profit margins for different actors in the chain?
- Market Access: How well are farmers connected to markets?
- Product Quality: What is the quality of the final product?
- Sustainability: What are the environmental impacts of the value chain?
- Farmer Income: How does the value chain affect farmers' incomes?
- Price Transmission: How quickly and accurately do price changes at one stage of the chain get transmitted to other stages?
- Waste Reduction: How much waste is generated at different stages of the chain?
- Traceability: How easily can products be traced back to their origin?

7. Challenges and Opportunities

- Challenges
  - Lack of Infrastructure: Poor roads, inadequate storage facilities, unreliable electricity supply.
  - Limited Access to Finance: Farmers and agri-enterprises often struggle to access credit.

  - Information Asymmetry: Unequal access to information about markets, prices, and technologies.
  - Market Volatility: Fluctuations in prices can make it difficult for farmers and agri-enterprises to plan.
  - Weak Governance: Lack of clear regulations and enforcement can create uncertainty and discourage investment.
  - Climate Change: Extreme weather events and changing climate patterns can disrupt agricultural production and value chains.
- Opportunities
  - Technology Adoption: Using technology to improve efficiency, productivity, and market access.
  - Value Addition: Processing and packaging agricultural products to increase their value.
  - Market Development: Expanding access to new markets, both domestic and international.
  - Collaboration: Building stronger relationships between actors in the value chain.
  - Sustainable Practices: Adopting sustainable farming practices to protect the environment and ensure long-term productivity.
  - Policy Support: Government policies that support agricultural development and value chain improvement.

8. Theoretical Frameworks

- New Institutional Economics: Examines the role of institutions (e.g., contracts, property rights) in shaping economic behavior within value chains.
- Transaction Cost Economics: Focuses on minimizing the costs of transactions between different actors in the value chain.
- Resource-Based View: Analyzes how firms can leverage their resources and capabilities to create a competitive advantage within the value chain.

Supply Chain Management: Applies principles of supply chain management to improve the efficiency and effectiveness of agricultural value chains.

In summary, studying agri-enterprises in service sectors within value chain models is essential for understanding how to improve agricultural systems. By analyzing the roles, relationships, and challenges of these enterprises, researchers and policymakers can identify opportunities to enhance efficiency, profitability, sustainability, and the livelihoods of farmers.

9

Study of Agri-Enterprises Involved in other Sectors (Value Chain Models)

Studying agri-enterprises involved in various industry sectors through Value Chain Models can provide insights into how agricultural products are grown, processed, marketed, and sold. Understanding these value chains is crucial for enhancing efficiency, increasing profitability, and improving sustainability in agricultural sectors. Below, we will discuss the concept of value chain models, how to analyzeagri-enterprises within these models, and provide examples.

Key Components of Value Chain Models

1. **Primary Production**: This involves the cultivation of crops or livestock production. It encompasses all activities related to planting, growing, harvesting, and maintaining the quality of the produce.
2. **Processing**: This stage includes transforming raw agricultural products into consumable goods or other forms suitable for market distribution. Processing can take many forms, from simple packaging to complex food manufacturing.
3. **Distribution**: This aspect covers the logistics of getting agricultural products from farms or processing plants to consumers. It includes transportation, warehousing, and inventory management.
4. **Marketing and Sales**: Effective marketing strategies are necessary for reaching consumers, which may involve branding, advertising, pricing strategies, and sales channels, including retail and online platforms.
5. **Consumption**: This is the final stage where consumers purchase and utilize the agricultural products. Understanding consumer behavior can influence the entire value chain.

Analyzing Agri-Enterprises Through Value Chain Models

1. **Mapping the Value Chain**: Identify all the critical activities within an agri-enterprise's value chain. Process mapping can be utilized to visualize the flow from production to consumption.

2. **Assessment of Value Addition**: Analyze where value is added at each stage of the chain. Determine cost structures, margins, and profitability to identify opportunities to increase value.
3. **Stakeholder Analysis**: Identify the key players in the value chain, such as farmers, cooperatives, processors, distributors, retailers, and customers. Understanding each player's role can help find leverage points for improvement.
4. **Supply Chain Integration**: Evaluate the integration between stages of production, processing, and distribution. Exploring vertical and horizontal integration strategies can streamline operations.
5. **Sustainability Considerations**: Assess environmental and social impacts at various stages of the value chain. Consider sustainable practices that can enhance long-term viability, such as organic farming, waste management, and ethical sourcing.
6. **Market Trends and Consumer Preferences**: Stay updated on market trends, technological advancements, and changes in consumer preferences that can affect the value chain. Innovations can open new avenues for value creation.

Examples of Value Chain Models in Agri-Enterprises

1. **Fruit and Vegetable Industry**
   - **Primary Production**: Farmers cultivate fruits and vegetables, leveraging modern agricultural practices.
   - **Processing**: Processing facilities can wash, cut, and package produce for retail or food service.
   - **Distribution**: Cold chain logistics ensure freshness during transportation to supermarkets.
   - **Marketing**: Branding efforts highlight local produce and organic options.
   - **Consumption**: Focus on health benefits and convenience in communication with consumers.
2. **Dairy Farming**
   - **Primary Production**: Dairy farmers manage the health and productivity of cattle.
   - **Processing**: Milk is pasteurized, packaged, and transformed into cheeses and yogurts.
   - **Distribution**: Established networks for transportation to retail outlets or direct sales.

- **Marketing**: Niche marketing for organic or artisan dairy products.
- **Consumption**: Analyzing demographic consumption trends for dairy products.

3. **Coffee Industry**
   - **Primary Production**: Coffee growers cultivate beans under specific climatic conditions.
   - **Processing**: The beans undergo drying, roasting, and grinding.
   - **Distribution**: E-commerce platforms for direct-to-consumer sales.
   - **Marketing**: Storytelling in branding emphasizes the source and ethical practices.
   - **Consumption**: Consumer preferences lean towards specialty coffee and sustainability.

Conclusion

The study of agri-enterprises through value chain models offers an analytical framework for improving efficiency, profitability, and sustainability in agriculture. By understanding how value is created and identifying opportunities within these chains, stakeholders can facilitate innovations, respond to market dynamics, and enhance overall agricultural productivity. This comprehensive approach fosters resilience within agri-enterprises and contributes to food security and economic development.

10

Study of Key Components of Value Chain in Agri-Enterprises

The study of agri-enterprises in the service sectors, particularly through the lens of value chain models, is crucial for understanding how agricultural products are transformed from raw materials into value-added goods and services. This involves analyzing the entire process of production, processing, distribution, and retailing, as well as the supporting services that enhance efficiency and effectiveness in these processes.

Key Components of Value Chains in Agri-Enterprises

1. **Input Supply**
   - **Description**: This segment includes all the services and products required to support agricultural production, such as seeds, fertilizers, machinery, and farming services (e.g., land preparation).
   - **Service Providers**: Agricultural suppliers, cooperatives, and input distribution companies.
2. **Production**
   - **Description**: The actual farming activities where crops are grown or livestock is reared. This stage can incorporate various services such as extension services, training, and digital solutions for precision agriculture.
   - **Key Considerations**: Sustainable practices, labor management, and crop yield optimization.
3. **Post-Harvest Handling and Processing**
   - **Description**: Activities that occur after harvesting, including cleaning, grading, packaging, and transforming raw products into processed foods.
   - **Value-Add Services**: Cold storage services, logistics, and supply chain management.

4. **Marketing and Distribution**
   - **Description**: The strategies and channels through which agricultural products are brought to market, including wholesalers, retailers, and direct selling platforms (e.g., farmer's markets, e-commerce).
   - **Role of Services**: Services such as market research, branding, advertising, and transportation play crucial roles in connecting producers to consumers.
5. **Consumption**
   - **Description**: The end use of agricultural products by consumers. Understanding consumer behavior and preferences is critical for agri-enterprises.
   - **Value-Added Services**: Nutrition education, food safety, and culinary services.
6. **Support Services**
   - **Description**: These include institutional support, financial services (microfinance, insurance), information technology, and logistics that influence the efficiency and effectiveness of the entire value chain.
   - **Importance**: They help in risk management, enhancing productivity, and improving market access.

Value Chain Models

1. **Linear Value Chain Model**: A simple representation where products flow from one stage to another. Useful for straightforward products but may overlook feedback loops and service components.
2. **Integrated Value Chain Model**: This model stresses the inter connectedness of different stages, highlighting the flow of information, inputs, and outputs. It emphasizes cooperation among different stakeholders.
3. **Dynamic Value Chain Model**: Focuses on adaptability and responsiveness to changes in market demand, technology, and regulations. This model can incorporate flexibility in responding to consumer preferences.
4. **Service-Dominant Logic (S-D Logic)**: A framework that prioritizes service provision and relationships over goods. It considers agricultural products as a service (value co-creation) where producers and consumers engage in interactive exchanges.

Case Studies and Applications

- **Agri-Tech Startups**: Companies that provide platforms for farmers to access market data, best practices, and customer connections through digital means.
- **Cooperatives**: Agri-cooperatives pooling resources for input supply, collective marketing, and shared services, thus reducing costs and risks.
- **Value Chains for Specialty Crops**: Focusing on high-value crops like organic produce or artisanal products that have niche markets and require specific handling.

Conclusion

Studying agri-enterprises in the service sectors through value chain models allows stakeholders to identify opportunities for innovation, efficiency improvement, and market expansion. By integrating service components into value chains, producers can enhance their competitiveness, respond effectively to consumer needs, and ensure sustainable practices in agriculture. The focus on value-added services is increasingly vital in a globalized economy where consumer preferences and market dynamics are continually evolving.

11

Learning About Concept of Project Formulation on Farming-Based Livelihood Systems Along with Cost and Profit Analysis

Project Formulation for Farming-Based Livelihood Systems

Project formulation is the process of systematically developing a project idea into a detailed and implementable project plan. It's like creating a roadmap that outlines what you want to achieve, how you'll achieve it, and what resources you'll need. For farming-based livelihoods, this process needs to be particularly sensitive to the unique challenges and opportunities of the agricultural sector.

Here's a step-by-step overview

1. Situation Analysis (Needs Assessment)

- Identify the Problem/Opportunity: What challenges are the target farmers facing? (e.g., low yields, lack of market access, climate change vulnerability, soil degradation, limited access to credit). Alternatively, what opportunities exist? (e.g., growing demand for organic produce, new technologies, government subsidies).
- Target Group Identification: Who will benefit from the project? Be specific. (e.g., smallholder farmers in a particular region, women farmers, youth). Understand their current farming practices, income levels, skills, and access to resources.
- Resource Assessment: What resources are available? (e.g., land, water, labor, existing infrastructure, local knowledge). Assess the quality and quantity of these resources.
- Stakeholder Analysis: Identify all stakeholders involved or affected by the project. (e.g., farmers, community leaders, government agencies, NGOs, input suppliers, buyers). Understand their interests, roles, and potential contributions.

- Environmental Assessment: How will the project impact the environment? Consider both positive and negative impacts. (e.g., water usage, soil erosion, biodiversity). This is often a preliminary assessment at this stage.

2. Goal and Objective Setting:

- Goal: The overall long-term impact you want to achieve. It should be broad and aspirational. (e.g., "To improve the livelihoods of smallholder farmers in [region]").
- Objectives: Specific, Measurable, Achievable, Relevant, and Time-bound (SMART) steps to achieve the goal. (e.g., "To increase average crop yields by 20% within 3 years," "To establish a farmer cooperative for marketing produce within 1 year," "To train 50 farmers in sustainable farming practices within 6 months").

3. Project Design

- Activities: Specific tasks that need to be carried out to achieve the objectives. (e.g., providing training, supplying seeds and fertilizers, constructing irrigation systems, organizing marketing events).
- Inputs: Resources required for the activities. (e.g., seeds, fertilizers, equipment, labor, training materials, consultants).
- Outputs: Tangible products or services that result from the activities. (e.g., trained farmers, increased crop production, established irrigation systems).
- Implementation Plan: A detailed schedule of activities, including timelines, responsibilities, and resource allocation. This often involves Gantt charts or similar project management tools.
- Management Structure: Define the roles and responsibilities of the project team. Who will be responsible for what? How will decisions be made?
- Risk Assessment: Identify potential risks that could hinder project success (e.g., drought, floods, market fluctuations, pest infestations, political instability). Develop mitigation strategies for each risk.

4. Cost and Profit Analysis (Economic and Financial Feasibility)

- This is a critical step, which we'll cover in detail below. It involves estimating the costs of the project and the expected revenues to determine if the project is financially viable.
- Consider different scenarios (best case, worst case, most likely case).

5. Monitoring and Evaluation (M&E)

- Indicators: Define how you will measure progress towards your objectives. (e.g., crop yields, farmer income, market prices, number of farmers trained).
- Data Collection Methods: How will you collect data on the indicators? (e.g., farmer surveys, crop cutting experiments, market price monitoring, training records).
- Reporting: How will you communicate project progress to stakeholders? (e.g., regular progress reports, meetings, site visits).
- Evaluation: Periodic assessments to determine if the project is achieving its objectives and to identify lessons learned. This could be mid-term and end-of-project evaluations.

6. Sustainability Plan

- How will the project's benefits continue after the project funding ends? (e.g., establishing farmer cooperatives, training local extension officers, developing market linkages).
- Consider environmental, social, and economic sustainability.

II. Cost and Profit Analysis for Farming Projects

Cost and profit analysis, also known as economic or financial feasibility analysis, is essential to determine whether a farming project is likely to be profitable and sustainable. It involves carefully estimating all the costs associated with the project and comparing them to the expected revenues.

A. Cost Estimation

- Fixed Costs (Overhead Costs): Costs that do not vary with the level of production. These costs are incurred regardless of whether you produce a lot or a little.
  - Land rent or mortgage payments
  - Depreciation of equipment (tractors, harvesters, irrigation systems)
  - Insurance
  - Property taxes
  - Administrative salaries
  - Interest on loans
- Variable Costs (Operating Costs): Costs that vary directly with the level of production. The more you produce, the higher these costs will be.
  - Seeds and seedlings
  - Fertilizers

  - Pesticides and herbicides
  - Water (irrigation costs)
  - Labor (planting, weeding, harvesting)
  - Fuel and electricity
  - Transportation costs
  - Marketing and sales expenses
  - Packaging materials
- Opportunity Cost: The value of the next best alternative that is forgone. For example, the income a farmer could earn from working in another job instead of farming their land. This is often considered in economic analysis but not always in financial analysis.
- Contingency Costs: A buffer added to the total cost estimate to account for unexpected expenses (e.g., price increases, weather-related losses). A common practice is to add 5-10% of the total cost as a contingency.
- Start-up Costs (Initial Investment): One-time costs incurred at the beginning of the project.
  - Land clearing and preparation
  - Purchase of equipment and machinery
  - Construction of buildings (storage, processing facilities)
  - Initial planting materials

B. Revenue Estimation

- Yield Prediction: Estimate the expected yield per unit area (e.g., kilograms per hectare). Consider factors like crop variety, soil quality, climate, and farming practices. Use historical data, research findings, and expert advice.
- Market Price: Research the current and expected market prices for the agricultural products. Consider factors like seasonality, demand, quality, and location.
- Sales Volume: Estimate the quantity of products you expect to sell. This will depend on your production capacity and market demand.
- Value-Added Products: If you plan to process or package your products, estimate the additional revenue you can generate.
- Government Subsidies and Incentives: Factor in any government payments or tax breaks that you are eligible for.
- Other Income: Consider any other sources of income, such as renting out equipment or providing agritourism services.

C. Profitability Analysis

- Gross Revenue: Total income from sales (Sales Volume x Market Price).
- Total Costs: Sum of all fixed costs and variable costs.
- Gross Profit: Gross Revenue - Variable Costs. This indicates the profitability of your core farming activities before considering overhead costs.
- Net Profit (Net Income): Gross Revenue - Total Costs. This is the bottom line – the profit remaining after all expenses have been paid.
- Profit Margin: (Net Profit / Gross Revenue) x 100. This indicates the percentage of revenue that translates into profit.
- Break-Even Analysis: Determines the production level (yield or sales volume) at which total revenue equals total costs. This helps you understand the minimum level of production needed to avoid losses.
    - Break-Even Point (in units) = Fixed Costs / (Selling Price per Unit - Variable Cost per Unit)
- Benefit-Cost Ratio (BCR): Present Value of Benefits / Present Value of Costs. A BCR greater than 1 indicates that the project is expected to generate more benefits than costs. This requires discounting future cash flows to their present value using an appropriate discount rate (which reflects the time value of money and the riskiness of the project).
- Return on Investment (ROI): (Net Profit / Total Investment) x 100. This indicates the percentage return on the money invested in the project.
- Payback Period: The time it takes for the project to generate enough cash flow to recover the initial investment.

D. Financial Statements

- Income Statement (Profit and Loss Statement): Summarizes the revenues, expenses, and profits of the project over a specific period (e.g., a year).
- Cash Flow Statement: Tracks the movement of cash into and out of the project over a specific period. This is crucial for managing liquidity and ensuring that you have enough cash to cover your expenses.
- Balance Sheet: A snapshot of the project's assets, liabilities, and equity at a specific point in time.

III. Important Considerations

- Data Quality: The accuracy of your cost and revenue estimates is critical. Use reliable data sources and consult with experts.
- Assumptions: Be aware of the assumptions you are making and how they could affect the results. Document your assumptions clearly.

- Sensitivity Analysis: Test how the profitability of the project changes when key variables (e.g., yield, price, costs) change. This helps you identify the most critical factors and assess the project's vulnerability.
- Time Value of Money: When analyzing projects that generate benefits over multiple years, it's important to consider the time value of money. Use discounting techniques to calculate the present value of future cash flows.
- Risk and Uncertainty: Agriculture is inherently risky due to factors like weather, pests, and market fluctuations. Incorporate risk assessment into your analysis and consider using risk management strategies like crop insurance or diversification.
- Software and Tools:Spreadsheet software (e.g., Microsoft Excel, Google Sheets) can be very helpful for performing cost and profit analysis. There are also specialized agricultural software packages available.
- Expert Advice: Don't hesitate to seek advice from agricultural economists, financial advisors, and experienced farmers.

Example Scenario

Let's say you are considering starting a 1-hectare tomato farm. You would need to estimate the following:

- Start-up Costs: Land preparation, seedlings, irrigation system.
- Fixed Costs: Land rent, depreciation of equipment.
- Variable Costs: Fertilizer, pesticides, labor, water.
- Yield: Expected kilograms of tomatoes per hectare.
- Market Price: Expected price per kilogram of tomatoes.

Then, you would use this information to calculate your gross revenue, total costs, net profit, profit margin, break-even point, and other profitability metrics. You could also perform a sensitivity analysis to see how your profit would change if the price of tomatoes decreased or if your yield was lower than expected.

In summary, project formulation is a systematic process of planning and designing a farming-based livelihood project, while cost and profit analysis is a critical component that determines the financial viability of the project. By carefully considering all the factors involved and using appropriate analytical tools, you can increase the chances of success and ensure that your farming project contributes to sustainable livelihoods. Remember to tailor your approach to the specific context and challenges of the target community.

12

Case Study of Startups in Agri-Sectors

I. Case Study Framework: Agri-Sector Startups

This framework focuses on analyzing the startup's value proposition, execution, and impact within the agricultural landscape.

A. Start-ups Profile

1. Name and Location: (e.g., "Agbotic, New York, USA")
2. Founding Date and Team: (Key founders, their backgrounds, and relevant experience in agriculture, technology, or business). Highlight the team's expertise and any advisory boards.
3. Mission and Vision: What problem are they solving? What is their long-term goal for transforming agriculture?
4. Legal Structure: (e.g., LLC, Corporation)
5. Funding History: Seed rounds, Series A, venture capital, grants, bootstrapping. Identify key investors and the amount raised.
6. Stage of Development: (e.g., Ideation, Seed, Early Growth, Scaling)

B. Value Proposition

1. Problem Addressed: Clearly define the specific challenge in agriculture that the startup is tackling. (e.g., labor shortages, inefficient irrigation, lack of market access for small farmers, food waste, soil degradation).
2. Solution Offered: Describe the startup's product or service. Be specific about how it works. (e.g., AI-powered yield prediction software, robotic harvesting systems, a mobile platform connecting farmers to buyers, a bio-fertilizer product).
3. Target Customer: Who benefits most from their solution? (e.g., Large-scale commercial farms, smallholder farmers, food processors, consumers, agricultural input suppliers).
4. Unique Selling Proposition (USP): What makes their solution different and better than existing alternatives? (e.g., Lower cost, higher accuracy, improved efficiency, sustainability benefits, enhanced data insights).

5. Value Creation: How does the solution create value for the customer (e.g., increased yields, reduced costs, improved quality, access to new markets, better resource management)? Quantify the value where possible.

C. Technology and Innovation

1. Core Technology: What technology is the startup using (e.g., AI, IoT, robotics, biotechnology, blockchain, sensors, drones, precision agriculture techniques)?
2. Innovation: How innovative is their approach? Is it a completely new technology, or an improvement on an existing one? Does it have patent protection or potential for it?
3. Data Usage: How does the startup collect, analyze, and use data? Is data privacy and security addressed?
4. Scalability: Can the technology be easily scaled to serve a larger customer base?
5. Integration: How well does the technology integrate with existing farming practices or systems?

D. Business Model and Operations

1. Revenue Model: How does the startup generate revenue? (e.g., Software-as-a-Service (SaaS) subscriptions, direct sales of products, transaction fees, data analytics services, consulting fees).
2. Distribution Channels: How does the startup reach its target customers? (e.g., Direct sales force, online marketplace, partnerships with agricultural retailers, distributors).
3. Marketing and Sales Strategy: How does the startup attract and acquire customers? (e.g., digital marketing, trade shows, farmer networks, pilot programs).
4. Operations: Describe the key operational activities of the startup. (e.g., software development, manufacturing, logistics, customer support, field operations).
5. Supply Chain: How does the startup manage its supply chain? (e.g., sourcing raw materials, manufacturing, distribution).
6. Key Partnerships: Does the startup rely on partnerships with other companies, research institutions, or government agencies? How important are these partnerships?

E. Market Analysis

1. Market Size and Opportunity: What is the size of the target market for their solution? Is the market growing?
2. Competitive Landscape: Who are the main competitors? What are their strengths and weaknesses? How does the startup differentiate itself?
3. Market Trends: What are the key trends shaping the agricultural industry? (e.g., increasing demand for sustainable agriculture, adoption of precision agriculture technologies, rise of e-commerce in agriculture).
4. Regulatory Environment: Are there any regulations that affect the startup's operations? (e.g., environmental regulations, food safety regulations, data privacy regulations).

F. Impact and Sustainability

1. Economic Impact: How does the startup contribute to economic growth in the agricultural sector? (e.g., creating jobs, increasing farm incomes, improving efficiency).
2. Social Impact: How does the startup address social challenges in agriculture? (e.g., improving food security, empowering smallholder farmers, promoting gender equality).
3. Environmental Impact: How does the startup contribute to environmental sustainability in agriculture? (e.g., reducing water consumption, minimizing pesticide use, improving soil health, reducing greenhouse gas emissions).
4. Sustainability of the Business Model: Is the startup's business model sustainable in the long term? Is it financially viable and environmentally responsible?

G. Challenges and Risks

1. Technical Risks: Are there any technical challenges associated with the startup's technology?
2. Market Risks: Are there any market risks that could affect the startup's success? (e.g., low adoption rates, competition from established players).
3. Financial Risks: Are there any financial risks that could threaten the startup's survival? (e.g., difficulty raising capital, high operating costs).
4. Regulatory Risks: Are there any regulatory risks that could impact the startup's operations?
5. Operational Risks: Are there any operational risks, such as supply chain disruptions?

H. Key Performance Indicators (KPIs)

1. Identify relevant KPIs to measure the startup's success. Examples:
    - Customer Acquisition Cost (CAC)
    - Customer Lifetime Value (CLTV)
    - Monthly Recurring Revenue (MRR) (if applicable)
    - Yield improvements for farmers using the technology
    - Water savings achieved
    - Reduction in pesticide use
    - Customer satisfaction scores
    - Market share
2. Conclusion and Recommendations
    - Summary of Findings: Summarize the key strengths and weaknesses of the start-ups.
    - Overall Assessment: Provide an overall assessment of the startup's potential for success.
    - Recommendations: Offer specific recommendations for how the startup could improve its business model, technology, or operations. Suggest potential areas for future research or development.
    - Lessons Learned: What broader lessons can be learned from this startup's experience that could benefit other agri-tech entrepreneurs?

II. Examples of Agri-Sector Startup Types (Potential Case Study Subjects)

Here are some categories of agri-startups, which might give you some ideas on who to study. You can often find startups in these areas through online searches, industry publications, and startup databases (like Crunchbase, PitchBook, etc.)

A. Precision Agriculture

1. Focus: Using technology (sensors, drones, data analytics) to optimize farming practices.
2. Examples: Companies that provide yield prediction, soil monitoring, automated irrigation, variable rate application of fertilizers/pesticides.

B. Agricultural Robotics

1. Focus: Developing robots for tasks like planting, weeding, harvesting, and livestock management.
2. Examples: Companies building robotic harvesters, autonomous tractors, and drone-based crop sprayers.

C. Vertical Farming

1. Focus: Growing crops indoors in stacked layers, often using hydroponics or aeroponics.
2. Examples: Companies operating vertical farms in urban areas, or developing technologies for controlled environment agriculture (CEA).

D. Agribusiness Marketplaces & Supply Chain Solutions

1. Focus: Connecting farmers directly with buyers, improving supply chain transparency, and reducing food waste.
2. Examples: Companies that operate online marketplaces for agricultural products, provide traceability solutions, or offer logistics services.

E. Biotechnology & Novel Farming Systems

1. Focus: Developing new crop varieties, bio-fertilizers, bio-pesticides, and sustainable farming practices.
2. Examples: Companies that use gene editing to improve crop yields, produce bio-based alternatives to synthetic fertilizers, or promote regenerative agriculture.

F. Farm Management Software

1. Focus: Providing software tools for farmers to manage their operations, track expenses, and make data-driven decisions.
2. Examples: Companies offering farm accounting software, crop planning tools, and livestock management systems.

G. Alternative Protein

1. Focus: Creating plant-based or cultivated (lab-grown) meat alternatives.
2. Examples: Companies focusing on myco protein production or cellular agriculture for beef, poultry, or seafood analogues.

III. Data Collection Methods

A. Publicly Available Information:

1. Company websites, press releases, news articles, social media profiles, investor presentations, Crunch base/Pitch book profiles.

B. Interviews

1. Ideally, interview the founders, employees, or customers of the startup. Prepare a structured interview guide in advance.

C. Site Visits

1. If possible, visit the startup's facilities or farms to observe their operations firsthand.

D. Industry Reports

1. Consult market research reports on the agricultural technology industry.
   - Important Considerations When Selecting a Startup:
   - Availability of Information: Choose a startup for which you can access sufficient information to conduct a thorough analysis.
   - Relevance: Select a startup that aligns with your interests and the focus of your study.
   - Impact: Consider the potential impact of the startup on the agricultural sector and society.
   - Stage: Early-stage startups are inherently riskier but can provide valuable insights into the challenges of building a new business. More established startups offer more data on their performance.

Agriculture has seen a surge in innovative start-ups aimed at addressing various issues within the sector. These agri-startups leverage technology and new business models to enhance productivity, sustainability, and efficiency in agricultural practices. Here's a case study overview highlighting some key areas of focus and successful examples of agri-startups:

13

Case Study Overview: Agri-Startups

1. Precision Agriculture

Example: Farmers Edge

Farmers Edge offers a comprehensive suite of solutions for farmers, utilizing data analytics, AI, and IoT (Internet of Things) technologies. The platform provides predictive analytics for crop health, soil quality, and weather, helping farmers make informed decisions.

- **Impact:** Improved yield and reduced resource wastage.
- **Technology:** Satellite imagery, data analytics, and machine learning.

2. Vertical Farming

Example: AeroFarms

AeroFarms uses aeroponic technology to grow crops without soil, using 95% less water than traditional farming methods. They operate indoor farms that can produce high yields all year round, independent of climate conditions.

- **Impact:** Increased food security in urban areas; fresh produce available year-round.
- **Technology:**Aeroponics, LED lighting, and climate control systems.

3. Supply Chain Optimization

Example: Example: AgroStar

AgroStar is a mobile platform connecting farmers with quality products and services, including fertilizers, pesticides, and seeds. They also provide guidance on best practices for crop management, enabling farmers to make better purchasing decisions.

- **Impact:** Streamlined supply chain and reduced costs for farmers.
- **Technology:** Mobile application, e-commerce, & educational resources.

4. Sustainable Agriculture

Example: Indigo Agriculture

Indigo focuses on harnessing nature's microbiomes to improve crop resilience and productivity. They work with farmers to apply beneficial microbes to crops, which enhances soil health and reduces the need for chemical fertilizers.

- **Impact:** Enhanced soil health and increased crop yields with sustainable practices.
- **Technology:** Microbial solutions and analytics.

5. Livestock Management

Example: Cowlar

Cowlar provides smart collars for livestock that monitor their health and reproductive cycles. The data helps farmers track the well-being of their animals and optimize breeding processes.

- **Impact:** Improved livestock health and productivity, reducing losses due to disease.
- **Technology:** Wearable technology, IoT, and data analytics.

6. Data-Driven Farming

Example: CropX

CropX offers soil sensors and data analytics to help farmers make informed irrigation decisions. The technology provides real-time data on soil moisture and other critical parameters, conserving water and improving crop health.

- **Impact:** Optimized irrigation schedules and resource management.
- **Technology:** Soil sensors, data analytics, and cloud computing.

Key Takeaways

- **Innovation in Agri-Tech:** Start-ups in the agri-sector utilize advanced technologies like AI, IoT, and biotechnology to increase efficiency and sustainability.
- **Focus on Sustainability:** Many agri-startups are developing solutions aimed at environmental sustainability, providing alternatives to traditional farming practices.
- **Empowering Farmers:** By improving access to products, data, and market information, these start-ups help farmers make better decisions, increasing both profitability and productivity.
- **Investment Opportunities:** The agri-tech sector is attracting significant investment, highlighting a growing recognition of the importance of innovation in agriculture.

Conclusion

The rise of agri-startups marks a critical shift toward technology-driven agriculture. These innovative companies are well-poised to tackle the challenges of food security, climate change, and sustainable farming practices, paving the way for a more resilient agrarian future. The success of these start-ups relies on their ability to adapt to the rapid technological evolution and changing farmer needs in an increasingly complex agricultural landscape.

14

Agri-Tech Startups Helping Farmers in India

India, with 118.7 million farmers, which accounts for more than half of its population, depends on agriculture as its primary source of income. However, the use of technology in the agricultural industry has been limited in India. As a result, the agriculture industry in India contributes merely 17-18% to its GDP. However, in the last couple of years, India has seen a rise in the number of agri-tech startups that are not only making technology more accessible but also helping these farmers to improve their lives. Today, we will take a look at top 10 agri-tech startups helping Indian farmers.

1. **CropIn Technology (2010):** Founded by Krishna Kumar and Kunal Prasad in 2010, CropIn provides one-stop SaaS-based solutions platform enabling agribusinesses to increase efficiency, productivity and sustainability. Its platform offers real-time weather updates, ability to manage farm activities, and predicts crop yields to minimize risk and yield better produce.
2. **Stellapps (2011):** Founded by RanjithMukundan, Ravishankar G. Shiroor, Praveen Nale, Ramakrishna Adukuri and VenkateshSeshasayee in 2011, Stellapps helps dairy farmers and cooperatives to maximize their profits by digitizing and optimizing Milk Procurement &Coldchain Management through its IoT-based SmartMoo platform.
3. **DeHaat (2012):** Founded by Shashank Kumar and Amrendra Singh in 2012, DeHaat offers agricultural services like access to agricultural inputs like seeds and fertilizers at affordable prices, personalized assistance, soil testing, weather reports, micro-finance, and insurance. DeHaat has raised $19.3 million to help Indian farmers.
4. **AgroStar (2013):** Founded by brothers ShardulSheth and SitanshuSheth in 2013, AgroStar offers an online marketplace for farmers to buy agricultural inputs. This agritech startup also helps farmers by providing real-time advice from experts on how to manage their crops and boost their yield. AgroStar has raised $42 million.

5. **EM3 Agri-Services (2013):** Founded by Rohtash Mal and his son Adwitya Mal in 2013, EM3 AgriServices helps small farmers who can't afford to buy expensive farming technology to rent specialization machines to boost productivity at cheaper costs.
6. **Aibono (2014):** Founded by VivekRajkumar in 2014, Aibono has pioneered the seed-to-plate concept, wherein it helps the farmers to increase their crop yield and also sell the produce to the retailers. The company uses soil sensors, IoT devices and imaging drones to collect farm data and transfers it onto their cloud platform which uses predictive analytics to help the farmers make smart agriculture decisions in order to increase their produce. Aibono has raised its budget $3.5 million.
7. **Ninjacart (2015):** Founded by ThirukumaranNagarajan, Kartheeswaran KK, AshutoshVikram, SharathLoganathan and VasudevanChinnathambi in June 2015, Ninjacart started as a B2C hyperlocal food delivery startup. It was transformed into a B2B agritechstartup in order to solve the fresh agricultural produce supply chain problem for the farmers and the retailers. So far, Ninjacart has raised $164.2 million from Tiger Global Management, Accel, Qualcomm Ventures and Steadview Capital.
8. **WayCool (2015):** Founded by KarthikJayaraman and Sanjay Dasari in July 2015, WayCool offers a B2B agritech platform that uses technology to control end-to-end agriculture supply chain right from farming inputs to last-mile distribution to help the farmers to sell their produce through multiple distribution channels. Till date, WayCool has raised $60.8 million
9. **Intello Labs (2016):** Founded by Milan Sharma, Nishant Mishra, Himani Shah and DevendraChandani in 2016, Intello Labs offers digital products like Intello Track, Intello Sort, Intello Pack and Intello Deep which uses technologies like computer vision and deep learning to help farmers, retailers and exporters to access the quality of their fruits and vegetables.
10. **Bijak (2019):** Founded by Nikhil Tripathi, DayaRai, JitenderBedwal, Mahesh Jakhotia and NukulUpadhye in 2019, Bijak enables traders and wholesalers to discover new suppliers, maintain ledgers, make payments, and access working capital through its app to scale their business. Bijak has raised $14.3 million.

Index

Feasibility 117, 122, 129, 136, 137, 139, 286, 287